Famous Planet Earth Caves Vol. 1

Sophie's Cave (Germany) - A Late Pleistocene Cave Bear Den

Authored By

Cajus G. Diedrich

PaleoLogic, Research Institute
Petra Bezruce 96, CZ-26751 Zdice
Czech Republic,
www.paleologic.eu

liability of Bentham Science Publishers shall be limited to the amount actually paid by you for the Work.

General:

1. Any dispute or claim arising out of or in connection with this License Agreement or the Work (including non-contractual disputes or claims) will be governed by and construed in accordance with the laws of the U.A.E. as applied in the Emirate of Dubai. Each party agrees that the courts of the Emirate of Dubai shall have exclusive jurisdiction to settle any dispute or claim arising out of or in connection with this License Agreement or the Work (including non-contractual disputes or claims).
2. Your rights under this License Agreement will automatically terminate without notice and without the need for a court order if at any point you breach any terms of this License Agreement. In no event will any delay or failure by Bentham Science Publishers in enforcing your compliance with this License Agreement constitute a waiver of any of its rights.
3. You acknowledge that you have read this License Agreement, and agree to be bound by its terms and conditions. To the extent that any other terms and conditions presented on any website of Bentham Science Publishers conflict with, or are inconsistent with, the terms and conditions set out in this License Agreement, you acknowledge that the terms and conditions set out in this License Agreement shall prevail.

Bentham Science Publishers Ltd.
Executive Suite Y - 2
PO Box 7917, Saif Zone
Sharjah, U.A.E.
subscriptions@benthamscience.org

CONTENTS

The Cover Image has been provided by Cajus G. Diedrich, the author of this book

FOREWORD

This book is focusing on a Late Pleistocene cave bear den in central Europe, which cave bear dens are larger cave systems, mostly filed up with ten thousands up to a half Million of cave bear bones. Herein, not only the bones of the Sophie's Cave in Upper Franconia, Bavaria southern Germany) are studied – it is the "den" and its change within 100.000 years and its interesting Wiesent valley sided position to the river terraces (or probably valley glaciers) within the Last Ice Age. These interdisciplinary sedimentological studies make the cave locally important for the geomorphology development of the past 5 Mio years and important to the questions of "glacial signs" in Upper Franconia. The cave bear research of the past 10 years has changed drastically the picture of "the cave bear" of the former Kurtén 1976 and Rabeder et al. 2000 cave bear books – which bears are indeed today splitted by DNA and osteometrical newest studies on skulls and teeth (= "cave bear clock") into several species and subspecies within the Late Pleistocene – the past 113.000 years. The cave bear ethology was for long misunderstood about European "cave bears", because all the extinct top predtors - steppe lions, Ice Age leopards, Ice Age spotted hyenas, and Ice Age wolves - the antagonists of cave bears, were not included into the "cave bear story". The predation and scavenging of cave bear carcass explained with perfect examples not only the Sophie's Cave cave bear bone taphonomy, and finally explained their deep hibernation in caves as protection against predation, such as the non-existence of "Neanderthal bone flutes", which were simply products of scavenging hyenas on cave bear cub hind leg bones. Whereas the large predators are few represented in the bone record, typical in Early/Middle Late Pleistocene middle high mountain boreal forests with nearly absence of the mammoth steppe game, quite unique in the European fossil record is a nearly complete Late Pleistocene weasel skeleton and den documentation. The herein presented Sophie's Cave and other caves of central Europe are international important furthermore due its contribution to the understanding of the "cave bear" exctinction, which can be demonstrated to be a chain reaction starting with climate change, boreal forest and food source disappearance up to the Last Glacial Maximum, and predation stress by Cromagnon humans or large carnivores. It is the first cave, where in Europe a Late Palaeolithic shamanic related reindeer antler/bone depot is proven explaining now the absence of "Ice Age cave art" in German and other western European caves. This Gravettian sanctuary falls within the main "cave bear" hunt period (Aurignacien-Gravettian) of the last and largest European cave bears.

W. Bleicher
Former Scientific Leader Heimatmuseum Schloss Hohenlimburg,
Alter Schlossweg 30, 58119 Hagen-Hohenlimburg,
Germany

PREFACE

The new Famous Planet Earth Caves e-book series challenges to present important caves all over the world scientifically, but somehow also popular in a mixture, that non-experts can understand their main importance. The Sophie's Cave in southern Germany belongs obviously to one of the oldest show caves in Germany and Europe and has a long Earth history starting 5 My ago with most importace for the Ice Age and its impact for the landscape reconstruction of northern Bavarian, Upper Frankonia. This is a beautiful dolomite rock and valley landscape, that became famous already in historic times. It was visited by many first and famous natural scientists from England (Buckland), France (Cuvier) and Germany (Goldfuss). The book is interdisciplinary but with a strong focus on its main importace – a Late Ice Age cave bear den within a former boreal forest environment of the medium high up to 550 a.s.l. elevated Franconia mountains. It presents a first detail scientific work after more then 150 years of non-research about the complex cave use by different cave bear species and subspecies, and their predation by lions, hyenas and wolves. It is completely different to the classical and old cave bear books, and updates much the knowledge about small and large cave bear ecology in Europe and their life and battle to survive in boreal forest mountain areas of central Europe, which was long misunderstood due to detail work lack about the top predators of Europe – the last lions and hyenas. Those specialized in mountain areas on cave bear feeding well to see at the Sophie's Cave bone material. The new systematic excavations are well documented, and are illustrated with the cave inhabiting or dwelling animals in action and "night-google vision cave view" – a new way of illustrating by a famous Ice Age artist, who painted also for the famous Beringeria Visitor Center in Youkon, Canada. From the archaeological view, the cave has another unique record, a larger reindeer antler and bone depot within the most nice speleothem decorated hall. Early modern Cromagnon humans of the Gravettian came into this cave-rich region for hunting migrating reindeeer herds in the steep valleys, but here in Upper Franconia, those did not paint or left engravings of animals mainly such as in Spanish and French caves. Here, and at other caves in Germany, shamanism was practiced by the Late Palaeolithic reindeer hunters similar as modern Scandinavian Sami people did until the Medieval times – depositions of antlers of their most important game, the reindeer at religious places. Finally the book gives a first insight about the typical modern cave animals, and postglacial use by Bronze Age, Iron Age and Medieval people, which resettled the valleys and cave entrances of Upper Franconia.

Cajus G. Diedrich
PaleoLogic, Research Institute, Petra Bezruce 96,
CZ-26751 Zdice, Czech Republic
Email: cdiedri@gmx.net

ACKNOWLEDGEMENTS

According to his unexpected race car accident in 2013, the book is dedicated to Mr. Wolfgang Dess. The project was supported by the cave owners Mr. W. and Mrs. S. Dess who financed the field work and science with their company Dess Grundstückverwaltungs GmbH & Co KG. Mrs. W. Wedewart gave logistic help. The cave guide Dr. T. Striebel, was responsible for further cave explorations and openings of cave branches, and supported with a new cave map ("cross section") and elevation point data. He finally curated the bones from the castle, which were rediscovered in 1998, and are now integrated in the collection of the museum. Thanks goes also to several junior and senior cavers from the speleoclubs Höhlengruppe DAV Erlangen, Speläogruppe Nordostoberfranken, Höhlenforschungsgruppe Blaustein (=HfgB) who helped in the bone cleaning and sediment sieving: S. and J. Uhl, I. and P. Heubes, N. Hedler, N. Niedling, M. and D. and K. Zistl., C. Moosdorf. M. and M. Conrad gave much help during the first and difficult "Hösch Cave" exploration and supported with cave equipment. M. Harder (speleogroup Forschergruppe Höhle und Karst Franken = FHKF) did the bat identification and supported with information on modern cave animals. Dr. H. Schabdach (also FHKF) gave personel information on the cave which allowed the rediscovery of three cave bear skulls in the Reindeer Hall; additionally, he was responsible over 15 years for the protection and study of the modern water organisms in the cave. Additonal thanks goes to Dr. H. Schabdach and R. Schoberth for the permit to use the preliminary basic cave maps. Futhermore, Mrs. R. Illmann (FHKF) send copies of the Staatsarchiv Bamberg from the years 1833-1837 concerning the historic conflicts between Graf zu Münster, Graf von Schönborn-Wiesentheit and the former inner ministery under King Ludwig. Dr. G. Schweigert from the Staatliche Museum für Naturkunde Stuttgart determined several fossils of the "Jurassic sponge reef fauna". PD Dr. G. Rösner supported with a photo of the lion jaw in the Bayerische Staatssammlung für Paläontologie und Geologie München. Pottery fragments were determined in their age and cultural relationship by Dr. W. Bleicher, who also gave information on shamanism about reindeer antler depots of the Oeger Cave (Sauerland), and from Russian historical antler deposits made by Sami. The two C^{14} analyses were thankfully made by D. Hood from the Beta Analytic Laboratory, Florida, USA (www.radiocarbon.com). Final thanks go to the biologist L.T. Parker for the time-consuming spell-check proof reading and helpful comments.

CONFLICT OF INTEREST

The author confirms that this chapter contents have no conflict of interest.

COLLECTIONS

The Jurassic fossils, such as all Pleistocene animal bones and archaeological finds found in 2011 are housed in the Rabenstein Castle Museum (= **MBR**). Two cave bear skulls, other cave bear bones, as well as bones from a horse and woolly rhinoceros are kept in the Urweltmuseum Oberfranken in Bayreuth (= **UM-O**). A lion's lower jaw is still present in the Bayrische Staatssammlung für Paläontologie der Universität München (= **BSP**), which was figured herein. The hyena jaw is stored in the British Museum of Natural History London (= **BMNHL**). A cave bear skull and some bones taken in the 80' ies from the cave were allowed to figure from the private Mr. H. Buchhaupt collection (= **BHC**).

CONTENT

One of the historically oldest discovered German show caves, the Sophie's Cave, in the Upper Franconia Mountains karst landscape of southern Germany (Bavaria) was opened in 1833 for visitors. The cave is reviewed in its history, geology and geomorphological valley development since Pliocene times, cave genesis and refill by sediments, and with many historical and new archaeological and palaeontological bone finds, especially cave bear remains. It is of European-scaled importance for the Late Palaeolithic cave archaeology, and Late Pleistocene palaeontology such as local climate and landscape change reconstructions. The cave supports to explain the European cave bear evolution and palaeoecology, especially the interactions between cave bears and their main antagonists – the largest top predators steppe lion prides or single leopards as bear killers deep in caves, or Ice Age spotted hyenas and wolf packs as scavengers and main carcass destructors and bone crushers. In Europe for the first time a documented weasel den was present which small carnivores used parts of the cave. Also quite rare are indirect proof (bite marks on cave bear bones) of Ice Age porcupines. Late Palaeolithic reindeer hunter humans (Gravettians) used finally a single central and by speleothems nice decorated hall area as sanctuary and deposited mainly large male shed reindeer antlers (over 100) for most probably shamanic rituals (antler C^{14}-dated: 30.833-30.340 cal. BP), which bone depot includes possibly even a mammoth pelvic (coxa C^{14}-dated: 29.340-28.600 cal. BP). Several thousands of years after a main cave ceiling collapse period and several passage or entrance blocking within the maximum glaciation (around 19.000 BP), Iron Age and Medieval humans finally used only anterior cave parts such as the yard in front of the cave and the lower cave parts. The cave is used modern by different cave dwelling or inhabiting animals (rarely bats), including small possibly endemic water related arthropods or annelids.

CHAPTER 1

INTRODUCTION

Abstract: The Sophie's Cave (Bavaria, Germany), is one of the oldest show caves in Germany, and even one of the famous Late Pleistocene cave bear den caves in Europe. The cave is situated with several other famous and larger cave bear den caves, such as the Zoolithen Cave and Große Teufels Cave (latter also show cave) in one of the most cave-rich regions of Germany, the Franconia Karst especially along the Wisent and Ahorn River valleys. The confusing names of the "Rabenstein Cave" (because of its close topographic situation to the Rabenstein Castle), and later named Sophie's Cave cave parts were renamed systematically. Along the historical wooden steps and trail several sherded ceramic lamps and pottery pieces were found, which are from the "oldest cave lighting system" known in a German show cave reaching back to its discovery in 1833. After the first report of the cave by the German priest Esper in 1774, several famous European natural scientists such as Rosenmüller, Goldfuss, Graf zu Münster, Buckland and Sternberg collected or started to describe first bone finds from the first discovered cave parts, which material went partly to Prague, Bayreuth or are lost. The youngest historical finds which seem to be in connection with the visitor show cave are few militaria from the King Ludwig I (1825-1848) regentship time.

Keywords: Sophies's Cave (Bavaria, Germany), large cave bear den Franconia Karst, show cave, cave discovery and exploration history since 1833, renaming of cave parts, oldest German cave lightening lamps preserved, militaria finds.

LATE PLEISTOCENE CAVE BEAR DENS IN UPPER FRANCONIA (SOUTHERN GERMANY)

Within the Late Pleistocene (116.000-24.000 BP), Millions of "cave bear" (*"Ursus spelaeus* Rosenmüller 1794" in former times) bones of about four different species and subspecies accumulated in many cave bear den caves of Europe, especially in lime-stone/dolimite rock and cave-rich regions within middle high elevated boreal forest conditions (Sauerland, Harz, Franconia, Swabian and Alpine Karst) [1 - 22]. The German localities are compiled herein in an overview map of important and larger Late Pleistocene cave bear dens of Germany (Fig. **1A**), including the herein described southern German Sophie's

Cave of Upper Franconia, Bavaria (Fig. **1B-C**).

Along the northern Bavarian Upper Franconia Wiesent and the adjacent rivers, more than 500 small caves and a few large ones were discovered [23 - 35], but few of them contain many cave bear or other in few amounts other megafauna bones (Fig. **1C**) [7, 22, 29]. Such Pleistocene fossil bones became known in this region by initial descriptions of the German priest Esper in 1774 due to first cave bear and other bone finds of the "great deluge times" [36] or first "hyena den cave" studies in the König-Ludwig's Cave (opposite Sophie's Cave, Fig. **1C**) by the English lord Buckland in 1823 [37]. In some of the Upper Franconia cave bear den caves, cave bear traces were found [38 - 40] in the form of cave bear polished corners, scratch marks on walls or hibernation nest depressions, which latter are named ichnologically *Ursalveolus carpathicus* Diedrich 2011 [41]. Cave bear footprints, named as *Ursichnus europaeus* Diedrich 2011, are not yet described from Upper Franconia or any other German caves. The best and most preserved long overlooked cave bear nests were mapped [7] and are described and figured herein for the Sophie's Cave. Bones accumulated partly over 100.000 years (and even longer since Middle Pleistocene Elsterian/Saalian [42], *e.g.* Zoolithen Cave and Zahnloch Cave) mainly in five of the larger Upper Franconia cave bear dens (Geisloch, Zoolithen, Sophie's, Große Teufels Caves, Genther Cave) (Fig. **1C**) [7]. These caves have each delivered several thousands of cave bear bones already starting with the "Spaten-forschung" in historic times [1, 2, 8, 21, 22, 25, 35 - 37, 42 - 44]. They still contain a high amount of bones including other rarer Late Pleistocene boreal forest and very few mammoth steppe valley migratory fauna animals (less than 3% in cave bear dens of Upper Franconia, imported into hyena/wolf dens) such as documented recently for the Sophie's [7, 45] and Zoolithen caves (Fig. **1C**) [42, 44, 46, 47]. Those eight cave bear dens are the most important ones in Upper Franconia, especially concerning the Late Pleistocene megafauna, which are clues for the European megafauna distribution including the cave bears, their ecology and evolution, or climate, geomorphology, and vegetation/glacier area reconstruction models [7]. From other caves (Neideck, Wunders, König-Ludwig caves), additional bone remains are known, but in much lesser amounts, which are often not well analysed yet.

At the Upper Franconia Zoolithen Cave, the first "*Ursus spelaeus* cave bear" holotype skull of Europe was discovered which was described first by Rosenmüller 1794 as an "extinct bear" [43, 47], it being recently revised to represent a skull of a subadult animal of the large cave bear "*Ursus ingressus*" [42]. The Zoolithen Cave bone taphonomy and several species were reviewed in the past five years with all its rediscovered Pleistocene lion, wolf, wolverine and hyena and cave bear holotype skulls [42, 44, 46, 47] which material is also important to understand the Sophie's Cave fauna and in larger scale boreal forest megafaunas of the last Ice Age in central Europe with cave bears being hunted and consumed by top predators, Ice Age wolves, Ice Age spotted hyenas and steppe lions [45].

This Zoolithen Cave is few kilometers southwest (Fig. **1C**) from the herein completely new explored, partly excavated, interdisciplinary analysed and by literature reviewed Sophie's Cave cave bear den (Fig. **1B**), of which some newer results have been published recently [7, 45]. In these Late Pleistocene middle high mountainous boreal forest region modern excavations, analyses of historical and new excavated bone material were missing for long. Also the large cave bear den Geisloch Cave (Fig. **1C**) is unstudied in a modern palaeontological/ sedimentological and cave bear find analyses [40], similar as the Große Teufels Cave, the second bone-rich show cave in Upper Franconia (Fig. **1C**). All other caves are partly unexplored or have delivered only few cave bear finds (*e.g.* Neideck Cave, Zahnloch Cave, Fig. **1C**) [7, 29].

After extensive historical "pickaxe and shovel excavations" or "cave bear bone hunts" in the Sophie's Cave and other Upper Franconia (especially Zoolithen Cave) [1, 2, 25, 35, 36, 37, 38, 42, 43] and several other European caves (e.g. Baumann's, Hermann's, Perick, Zoolithen, Mixnitz, Sloup, Sophie's caves) [1 - 22] "cave bear skeletons" were mounted in many show caves and several European museums, in all cases incorrectly or incomplete. It is known today, that those are composed of different cave bear individuals, age classes, sex and even subspecies [e.g. 14, 42] such as found still today in the Große Teufels Cave cave bear composite skeleton. Herein, for the first time, a small cave bear skeleton is arranged from the Sophie's Cave material anatomically nearly complete and correct at least using adult small cave bear subspecies material, that has included unique to Europe added to a skeleton, the nine "tongue bones" (see Chapter 3).

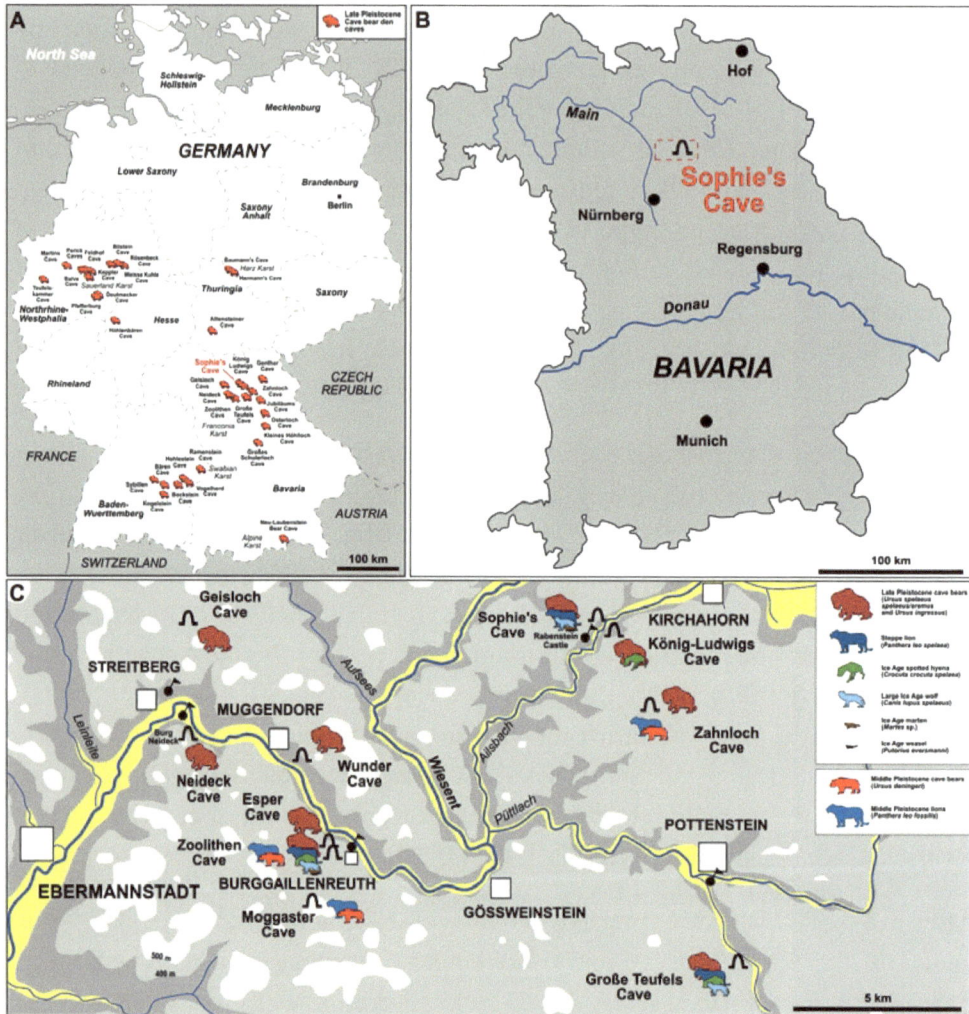

Figure 1. A. Main important cave bear den sites in Germany (composed from [1 - 22]). **B.** Position of the Sophie's Cave in northern Bavaria, southern Germany. **C.** Important Late Pleistocene cave sites and cave bear/hyena/wolf and small carnivore dens along the Wiesent River and adjacent streams (modified after [7] with adds of Middle Pleistocene cave bear den sites).

Those different small (*e.g.Ursus spelaeus* eremus/spelaeus Rabeder *et al*. 2004) [17, 20] and "largest" (*Ursus ingressus* Rabeder *et al*. 2004) [17, 20] partly DNA tested (*e.g.* Zoolithen Cave) [16] cave bear species and subspecies from the Late Pleistocene [15 - 20] are recently known for Upper Franconia since first new studies at the Sophie's [7] and Zoolithen [42] caves, and the herein presented osteologiccal bone atlas details (see Chapter 3). At least two species/subspecies seem to have populated even in coexistence the boreal forests in the early/middle Late Pleistocene of Germany [14, 15 - 20, 42] including Upper Franconia. The "cave bear clocks" demonstrated on cave bear tooth material for the Sophie's and Zoolithen caves [7, 42] furthermore a partly different biostratigraphic occurrence and non-overlap (stratigraphic occurrences after each others: first smaller *U. spelaeus* subsp., and above *U. ingressus*) of some subspecies/species as known for southern German Swabian cave bear dens [15]. The cave bear bonebed layers in the Sophie's Cave are partly completely autochthonous in some branches, which makes them valuable for future absolute dating and evolutionary, such as taphonomic and biostratigraphic analyses. This situation is completely different in the Zoolithen Cave, where final Late Pleistocene floods mixed all Late Middle (Saalian) to Latest Late Pleistocene (Weichselian) layers within a single bonebed, similar as found at the Zahnloch Cave [42].

THE SOPHIE'S CAVE CAVE BEAR DEN BESIDES RABENSTEIN CASTLE

This cave - a classical Late Pleistocene cave bear den – is situated in the North Bavarian Franconia Karst of southern Germany (Central Europe, Fig. **1B**). The 900 m long cave [34, 48] is positioned at the northwestern valley slope of the narrow and meandering Aisbach Valley (375 m a.s.l.). It is close to the small village of Kirchahorn in the Ahorntal (MTB 1:25 No. 6134 Waischenfeld) in the Upper Franconia Landkreis Bayreuth, very close to the Rabenstein Castle (Fig. **2**, GPS coordinates Google Earth: 49°49'37.32" North, 11°22'31.79" East). The 18 Meter wide, six meters high cave entrance of the Sophie's Cave is about 411 meters a.s.l, directly below the 443 meter high elevated Clausstein Chapel (Fig. **2**) [34, 48]. Today this cave (internet: http://de.wikipedia.org /wiki/Sophienh%C3%B6hle) is besides the larger Große Teufels Cave in the region open to visitors. The cave system consists of several halls, chambers, and passages on four different elevation levels, which are connected to each others (Fig. **3**). The show cave part is 200 meters long (Fig. **3**) [48]. From 1833 starting, and within the time of the first researchers (Fig. **4**), visitors were guided into the cave along wooden steps and constructions first with torches [48] and then with the first "cave lighting" (Fig. **5**). A first electric lighting system remained until 1971, when it was changed to new electrical lamps, with further changes in 2001. Finally in 2012 installed, it is in standard for show caves with "cold lights" to avoid the growth of bacteria, mushrooms and algae, especially on bones and speleothems.

Rabenstein Castle   **Clausstein Chapel**

Sophie's Cave entrance

Figure 2. Rabenstein Castle, Clausstein Chapel and entrance of the Sophie's Cave near Kirchahorn in the Ahorn Valley (Upper Franconia, Germany).

Sophie's Cave Discovery and History

The discovery and exploring history of different parts of the Sophie's Cave by different famous German historical researchers of the 19th century (Fig. **4**) can be reconstructed using a. the few historical published reports [48 - 51] and b. the new exploring/excavation results. Unpublished information comes from archive transcripts of law conflicts [52 - 54], personal communications of regional cave club members (speleogroup Forscher-gruppe Höhle und Karst Franken = FHKF), and interviews with former cave guides (*e.g.* Mr. Hoesch). The main historical "events" are listed herein chronologically using the above mentioned information:

In the 11th-13th Centuries already in the Early Medieval times, the first hall was named "Ahornloch" (today = Ahornloch Hall), after the knight "Von Ahorn". The former fortification of the Ahorn Castle (which was a wooden fortification) is today located under the Clausstein Chappel fundaments, 30 meters above the Sophie's Cave entrance. The castle inhabitants must have used the anterior accessible parts of the Sophie's Cave, and also the "Hösch Cave" (= Hösch Chambers), whose entrance is situated below the Sophie's Cave entrance (Fig. **3**).

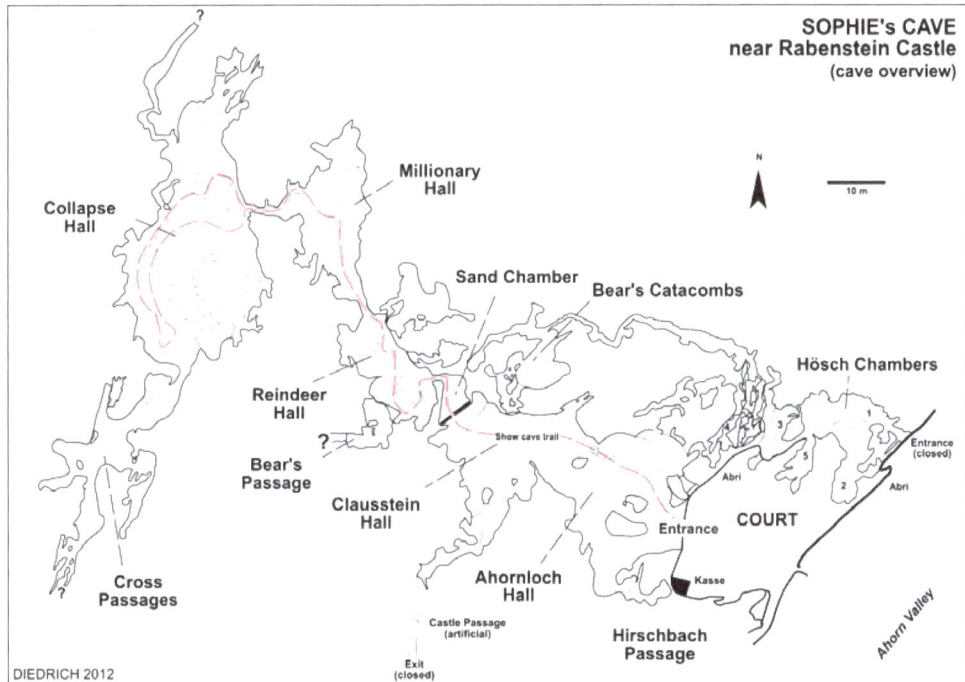

Figure 3. Names of parts of the Sophie's Cave (cave map outlines after unpublished map of Schoberth and Schabdach).

In the year 1490, the "Ahornloch", the first large hall (entrance hall) was mentioned in literature for the first time [48]. Later, the "Clauststein Cave" (today = Clausstein Hall) was briefly described. The first documentary evidence followed as a result of mining experiments to obtain "phosphatic cave sediments" for black powder production purposes. This was done by H. Breu from Bayreuth [48].

In 1774 the Uttenreuth priest, Johann Friedrich Esper (Fig. **4**) published the first palaeontological finds from the most famous bone–rich cave, "Zoolithen Cave" near Burggaillenreuth, which is situated only 12 km from the Sophie's Cave (Zoolithen = "fossilized bones") [36]. He also mentioned the first "Zoolithen" from the "Rabensteinhöhle" [36], which is synonymous with the "Ahornloch Cave", but he confused this cave with the opposite large cave entrance of the "Kühloch Cave" [37], which today is called "König-Ludwigs Cave" (Fig. **4**). After pickaxe excavations in 1788 in the "Ahornloch", a new hall was discovered and named "Clausstein Cave" [49] (today = Clausstein Hall). After the discovery, more extensive "treasure hunting" took place. During several digs, the shovel and pickaxe work didn't give attention to ceramic fragments or smaller bones which can be found today in front of the cave, also in a diagonal shaft of the Ahornloch Hall (which was refilled in modern history).

Figure 4. The first explorers of "Zoolithen" (= bone fossils) of the Sophie's Cave and visitors were from left to right: the priest J.F. Esper (1732-1781), the anatom J.C. Rosenmüller (1771-1820), the zoologist G.A. Goldfuss (1782-1848) and under König-Ludwig 1 (1786-1868) hired governor of Bayreuth Graf Georg zu Münster (1776-1844) – which were the pioneers of a new science, the "Palaeontology" (from [35, 36, 43]).

The pioneer Ice Age Palaeontologist Rosenmüller (Fig. **4**) did most of his excavations before 1804 and published his research about in the Zoolithen Cave near Burggail-lenreuth. He studied especially the cave bears and created the holotype skull by the binominal zoological name of "*Ursus spelaeus* Rosenmüller 1794". He also did some research on material from the Rabenstein Cave (=

Sophie's Cave) which was mentioned little in his work [43].

A first cave guide for the Franconia caves was published by the most famous German Palaeontologist Goldfuss in 1810 (Fig. **4**). From the Sophie's Cave described "gold sands" and cave "bear bones" (sadly without any figures) which he found in the "Rabenstein Cave" – as known now to be the first hall of the Sophie's Cave [35]. In his work, he named binominal the first "cave lion *Felis spelaea* Goldfuss 1810" using a skull from the Zoolithen Cave, which today it is known as the steppe lion *Panthera leo spelaea* (Goldfuss 1810) [44]. In his historical hiking guide book J. Heller decribed 1829 [2] the first "bone-rich" cave halls of the Sophie's Cave, although he himself never worked there.

In 1833 the former owner of the Rabenstein Castle and the Sophie's Cave, Count Erwein von Schönborn-Wiesentheit gave orders to his castle gardener M. Koch to extend the cave by removing further sediments [48]. To honor his wife, the Countess Sophie von Schönborn, born to Eltz, the daugther in law of the castle owner, the newly discovered part of the cave system (starting behind today's Sand-Halle) was named "Sophie's Cave" [48]. Schönborn gave instructions to install wooden trails and steps within the new part, which started in the Reindeer Hall, ending in the Millionary Hall (possibly figured by Rothbarth in 1850, Fig. **5A**) [48].

The cave was opened 1834 to the first tourists. It seems, the oldest historical lighting by ceramic oil lamps or reused old pottery was already installed at the beginning along the historical visitor trail loop (Fig. **5B**). With the discovery of the new parts of the cave system behind the Reindeer Hall, a law conflict arose between the Grafen zu Schönborn and the Grafen Georg zu Münster (Fig. **4**). Graf zu Münster wanted to obtain some of the "Zoolithen", especially skulls and some bones for the Bavarian Naturalienkabinett Bayreuth. This attempt failed although he reported 20 cave bear skulls, a mammoth pelvic, and dozens of reindeer antlers in Reindeer Hall [52 - 53]. Instead, Graf zu Schönborn exposed several cave bear skulls, a wolf, and possibly a hyena skull, and some other bones, which came to daylight from the cave extension works by his gardener. All the material which was stored in the castle (possibly until 1975) seems to have been lost. Some of the remains discovered by Graf zu Münster might be preserved *in situ* in the Reindeer Hall.

K. Sternberg reported in 1835 in front of the Natural History Society of Prague [51] about cave bear skulls, lower jaws from lions and hyenas, a mammoth pelvic bone and several reindeer antlers after his cave visit. He mentioned the construction of a museum in the Rabenstein Castle by some "Erlanger Professor". Meanwhile, several of Graf zu Münster's reported specimens and later lost objects seem to have been exposed in the castle quite early, whereas some material was unmentioned transferred (after own rediscoveries in their collection) into the today's Nationalmuseum Prague.

In 1837 C. Hösch discovered another cave part from the large cave system, not far and below the today's entrance, which was named "Hösch Cave". In the third hall he took a piece of "prehistoric pottery" and gave it to "Bayreuth". Also with this discovery, another law conflict arose [54] with the result of an immediate closing of the new discovered entrance. Since that time, the entrance of the Hösch Chambers is blocked by a large rock gravel cone [54].

Side branches of today's Collapse Hall seem not to have been opened before 1833, although the oldest graffiti inscriptions at the end of the Cross Passages date to the year 1853.

Neischl did new excavations in 1905, mainly in the herein new named Bear's Passage [25], which was accessible already *via* the Reindeer Hall. His material was used for a cave bear composite skeleton which received somehow the name "Benno". The skeleton was made as known from the new research in composition of a small cave bear species (*U. s. eremus/spelaeus*) but also the larger (*U. ingressus*), which bones originated from different cave areas and older/younger stratigraphic positions.

In the year 1975, several skulls and large bones were removed without permit from the extremely difficult to access Hösch Chambers. These bones were taken by speleologists of the cave club Forschergruppe der Höhlen- und Karstforscher Franken and were "given" to the city of Pottenstein. This again caused another legal conflict. About seven skulls and other materials were moved finally back to the former Rabenstein Castle owner, but are lost today.

H. Schabdach compiled in 1998, a first booklet the historical text passages describing the cave and bones. He added in this first small cave guide book a coarse 3D cave model and figured mainly speleothem photos, but also the modern cave animals without a scientific analyses of the cave or its contents [48].

In 2001 the cave trails and lighting systems were renewed.

L. Steguweit made in 2007 an archaeological trench in front of the cave, but too close to the rockshelter-like walls. He found only a few Prehistoric, Medieval and Modern ceramic and bone remains [55].

The New Cave Research

The Sophie's Cave and the courtyard in front of the entrance were surveyed in 2011 for the first time since 1833 interdisciplinary concerning its history, archaeology, geology, and palaeontology resulting in climatic and geomorphological influenced cave use by animals and humans [7]. The results are presented herein in all details and further unpublished material and comprise also discovered cave bear fossils which were placed in "bone dumps" (missing skulls, jaws and teeth = souvenirs) in the cave, or material which is stored in different museums/private collections. About 2,000 bones had been rescued in 2011 from those historic dumps from two main cave areas, Bear's Passage and Bear's Catacombs, which allowed reconstruction of an anatomically perfect and nearly complete small cave bear type skeleton (see Chapter 3). Also several important pottery and bone remains have been discovered in the Reindeer Hall, such as a mammoth coxa, and several hidden mud-covered reindeer antlers (see Chapter 4). Unexpectedly, a larger amount of phosphatic wolf excrements were sieved from the historically reworked sediments of the Bear's Passage, such as many weasel remains including a skeleton remain with its skull.

Whereas in all Upper Franconia caves not one systematic cave bear excavation was made since 1794 (and even before), and the "rescue excavations" at the Zoolithen Cave in the 80's used similar historical pickaxe methods (made by University Erlangen/FHKF) [42]. Those methods resulted also from to the general non-protection of any fossils in Bavaria. First new systematic surface excavations in Upper Franconia dealing mainly with cave bear bonebeds are presented herein.

The book presents the complete known Sophie's Cave macrofauna and different cave bear species/subspecies in a kind of a monograph. It is the first book presenting a typical European "cave bear den", the life, death and relationship between cave bears (and different species/subspecies) and their top antagonists - predators and scavengers (lions, hyenas, wolves). It is chaptered after "time periods" to understand the geomorphological change since Pliocene times and climatic and river terrace impact onto the "cave faunas"- especially cave bear presence and extinction.

Renaming of the Confusing Cave Part Names

The cave itself and chamber/passage parts of the Sophie's Cave received several names. In the cave cadastre the complete system has the name "Clausstein Cave complex" (cadastre no. B27) including now also the lower part of the "Hösch Cave" (no. B24) [48]). The Sophie's Cave should have been named the "Ahorn Cave" or "Rabenstein Cave", because those are the first written historic names [36, 37, 49]. For touristic reasons and "traditions", the cave system has long been called Sophie's Cave. Only the posterior part starting after the today called Sand Chamber was initially called Sophie's Cave. The first to third Sophies's Cave halls were named in 1833 "Erste bis Dritte Abtheilung" [48]. Some historical names were reused for naming the cave, but those were modified in a systematic conventional way after a. halls, b. chambers and their connecting c. passages, which form different parts of the 900 meter [48] long Sophie's Cave (Fig. **3**).

Ahornloch Hall: Until the 15th centuries, this first hall was named "Ahornloch" after the noble von Ahorn. Originally, there was a wooden fortification and small castle (11-13th Centuries) but it has been overbuilt by the Clausstein Chapel (Fig. **2**). Also the entrance area of the Hösch Chambers was used in the Early Medieval times. Now, only the first large hall is called "Ahornloch Hall".

Hirschbach Passage: This side-branch of the Ahornloch Hall was historically named "Hirschbach Cave".

Clausstein Hall: First it was named "Clausstein Cave", after the Clausstein Chapel. From this cave area, Bronze Age tools and Iron Age pottery fragments are mentioned, such as cave bear remains, which were found in 1832 during the cave extension visitor loop by the castle gardener. The sediment from this hall was – what we know today – shoveled into two more diagonal, deeper connected halls which were heavily refilled until 1837.

Sand Chamber: This herein new named part of the cave is between the Clausstein Hall and Reindeer Hall. It was the key chamber for the discovery of the posterior larger parts of the cave system which were opened in 1833. The chamber is filled with sand and clay deposits, and received its name after the "golden dolomitic ash sands".

Reinder Hall: 1833 the most speleothem-rich hall was first named "Erste Abteilung", which is revised herein after the important reindeer antler finds.

Bear's Passage: A side passage of the Reindeer Hall which is "full of cave bear bones" and other Pleistocene animal remains, was formerly named "Bears Cave". This passage seems to be connected with the blocked former cave bear den entrance. In 1905, the distal part of the passage, which can be reached only by ladders about three meters up from the Reindeer Hall floor, was partly excavated, leaving thereby a large bone dump without selected skulls, jaws and teeth. From this passage, the "Benno" cave bear skeleton skull was taken. This was proven because in 2011 new sieved and rediscovered teeth of this skull (canine and I3) fit exactly into the alveols of the skull. From this passage in the today's new skeleton composite are integrated pelvic, many vertebrae, ribs, and leg bones, which must have been found historically. Those were partly also used in the first composite reconstruction.

Millionary Hall: This area was named in 1833 "Zweite Abteilung" and is renamed herein after the largest speleothem stalagmite "Millionary".

Collapse Hall: The largest hall in the cave system was named in 1833 "Dritte Abteilung". It was given the new name after the massive collapse of its ceiling.

A
B
SOPHIE's CAVE
near Rabenstein Castle
(Modern historic lighting finds)
N
10 m
Lighting ceramics
Millionary Hall
Bulk vessel bottom
Lamp 3
Bulk vessel partly damaged
Slate pencil 2
Bullet cartridge
Cooking pot
Lamp 1
Lamp 2
Military helmet golden oak leaf decoration
1833
1833
1853
1837
Reindeer Hall
Cross Passages
Slate pencil 1
Lamp 4 (missing)
?
Sand Chamber
Clausstein Hall
Oldest Grafitti
Historic show cave trail
Abri
Abri
Entrance (closed)
Entrance
COURT
Cashier
Ahornloch Hall
Burg passage (artificial)
Exit (closed)
Hirschbach Passage
Ahorn Valley
DIEDRICH 2012

Figure 5. A. Historical clay lithography demonstrating first torch lighting during visits in the Reindeer Hall (from [48]), **B.** Overview of the Modern times cave lighting and militaria finds in the Sophie's Cave (after 1833) and explorer history years of the individual cave parts after cave grafitti and documents (cave map outlines after unpublished map of Schobert an Schabdach).

Cross Passages: These are the most southwestern passages, which are connected to the Collapse Hall. They seem to continue, but are actually filled up partly up to the ceiling with fluvial sediments.

Hösch Chambers: Initially as "Hösch Cave" (no. B24) named by the miller C. Hösch, the discovered part was quickly closed by a gravel dump cone by officials in 1837. The entrance, on a rock shelter-like cliff, is about 10 meters deeper than the modern Sophie's Cave visitor entrance. Not far beside, another entrance was kept open, but today it is sealed by a locked metal door. In former times the name "Vierte Abteilung" was used for the blocked diagonal passage between "Ahornloch Hall" and "Hösch Chambers". The first three medium-sized chambers were subdivided into four rooms by C. Hösch, with numbers 1 to 4, whereas a side-branch received a new number herein with chamber 5. In the fourth chamber, Hösch removed the historically mentioned prehistoric pottery, which could not be relocated in the museum in Bayreuth anymore. Those chambers are the main, non-disturbed "archaeological part" of the cave system, whereas Bronze Age, Iron Age and mainly Medieval ceramic fragments *in situ* (or dug in modern times by badger/fox) indicate future possible finds.

Bear's Catacombs: The last chamber and its branches are called Bear's Catacombs because of the abundance of cave bear bone remains from the large *U. ingressus* species. There is one small chamber that is connected to very small unexcavated passages in the Clausstein Hall, but their connections are blocked by historical to modern shovel activities.

The Oldest Preserved Show Cave Lighting System Ceramic Lamps in Germany

With the removal of modern historically imported and accumulated mud (since 1833), and during the cleaning of several sinter terrace basins and other cave parts in the Sand Chamber, Reindeer, and Millionary Halls, historical ceramic pieces of oil lamps were found. Several pottery fragments were fitted to complete or nearly complete oil lamps (Fig. **6**). Some slate pencils and a few pieces of militaria from the 19th century have been found in the upper modern times layers or on rock surfaces or hidden in a niche behind the metal door. Oil lamp fragments and smashed ceramics were also found along the entire historic (similar to modern) cave trail up to the Collapse Hall (or old name "3 Abteilung"). The incomplete pottery or complete oil lamps are in most cases from the first lighting system (= deposited lamps on larger rocks, or in niches) along the trail and the speleothems from 1833 (Fig. **5B**). All ceramic lamps and reused incomplete pottery parts (Fig. **6**) were found best preserved in the Sand Chamber, Reinder, and Millionary Halls, but a few remains were found also with few fragments in the last Collapse Hall which was the end of the main visitor trail. It also seems that the large Collapse Hall was opened a few years later with changes in the trail.

The historic reused "kitchen pottery" [56] is dated by the oldest reused bulk vassle/handle pot that was found smashed in pieces near the Millionary stalagmite in a historic made depression (Figs. **5B, 6A**) which pot form can be dated well around 1800 by comparison of a "kitchen pottery inventory of southern Germany" [57]. It is possibly older and was reused for lighting purposes. It was definitely imported into the cave incomplete because sediment sievings did not result in the finds of parts of one side. This missing part must have been smashed outside the cave in order to make a large "ceramic oil lamp". The find position coincides with the largest and most attractive stalagmite, in the deeper part of the cave, which had to be lighted with the largest lamp. The oldest known preserved show cave lighting system of Germany was the result of the orders of Graf zu Schönborn, who did not allow visitation of the cave with the use of torches (except magnesium torches), such torch visits were illustrated in 1850 (Fig. **5A**). The use of ceramic oil lamps were also mentioned, but not illustrated with such finds for other show caves of Europe, whereas the lamps changed later to iron lamps (*e.g.* from mining) [58, 59].

Handle, Halfshelled Ceramic Oil Lamps are in some cases complete (or nearly complete) after the gluing of several fragments. There are two more or less complete specimens (Figs. **6D, E**). Ceramic pieces of this oil lamp type were found in three cave areas beside and on the trails, which indicates the presence of at least six more oil lamps of the same style (maybe eight in total). Those pieces on the floor, on rocks or niches positioned along the wooden trail indicate where oil lamps lightened the visitor loops. Two nearly complete lamps were found in the Sand Chamber (Figs. **5B**, **6E**) in some niche behind the modern wall, and also in the Millionary Hall (Figs. **5B**, **6D**), deposited on the first large dropped ceiling block in the centre of the hall. In those, such as in the Reindeer Hall, further single non-fitting oil lamp fragments were collected. Also beside the entrance holes of the Bear's Passage, a complete lamp was found and taken in the 90's (Schabdach pers. com. who thought it to be of Medieval or older times), but is lost now.

Pressed Ceramic Oil Lamps are represented by only one unbroken complete piece which was discovered in a niche behind the wall of the Sand Chamber (Fig. **6F**).

A **Ceramic Henkel Pot** was shattered and the pieces were found in a larger niche opposite the large Millonary stalagmite (Fig. **5B**). This pot must have been opened outside the cave, possibly to obtain a similar, but much larger, oil lamp form. The decorations on the pots exterior consist of three linear horizontal dark-red color stripes on the schoulder area and orange coloured glazing on the inner side only (Fig. **6A**). It can be dated by comparison to south German Bavarian figurations of very similar pottery forms [57, 60], especially henkel pot variants from a basement treasure find (hundreds of pottery pieces) from the city of Schwäbisch Gmünd, around 1800 [57]. The henkel pot from the Sophie's Cave seems to have been reused in 1833 as a "large oil lamp". The fact it was used to burn materials is indicated by the black burned signs on the outside of the pot (Fig. **6A**). Another similar, but on both sides glazed pot base part was found in several pieces in the corner of the small Millionary stalagmite (Fig. **5B**, **6C**). The pot was formatted again outside the cave to obtain another "oil lamp form" with the pot basement only. This two-sided brown-colour glazed pot fragment seems to be younger than the first one, and must originate due to double-size glazing from the 19th century.

A strongly incomplete ceramic **Cooking Pot** was possibly a casserole pot [56, 57] (Fig. **6B**). It was also discovered with several pieces covered by dirt behind the Sand Chamber wall (Fig. **5B**). It was found with another incomplete ceramic cup. The pot dates into the beginning of the 19th century [56, 57].

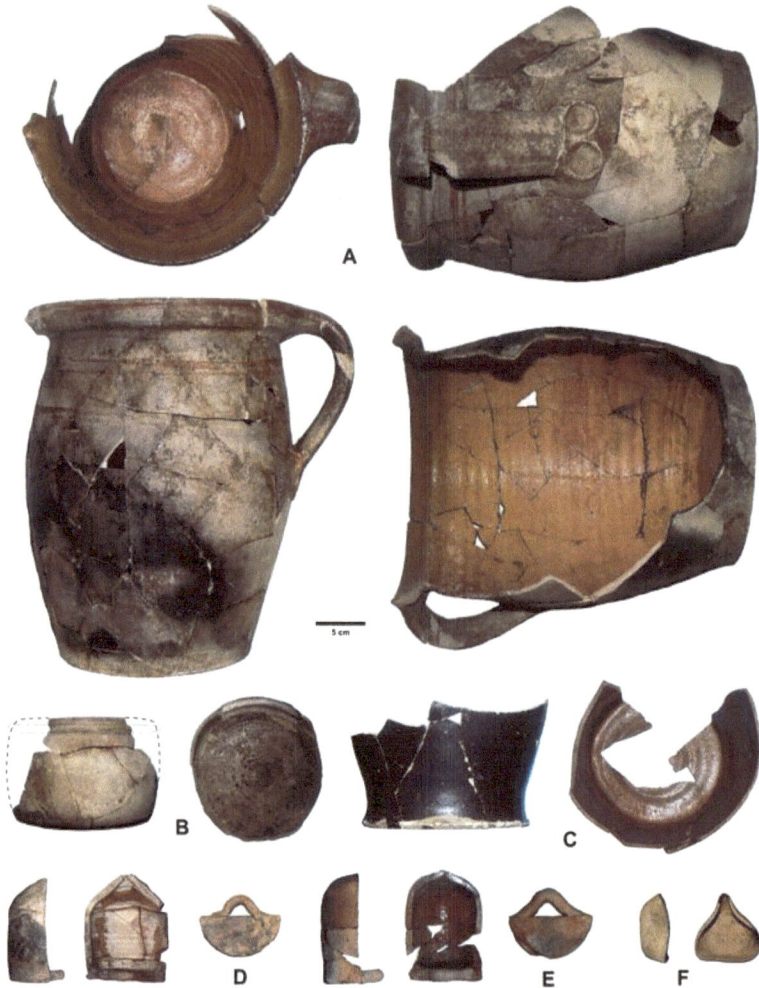

Figure 6. The oldest European show cave historic lighting ceramics (partly recycled cooking pots, and bulk vessels of the 18th and 19th Centuries) such as ceramic oil lamps from the Sophie's Cave. **A.** Handle pot with inner glazing around 1800, reused but partly smashed and opened to use as large oil lamp opposite the large Millionary stalagmite. **B.** Incomplete cooking pot (?casserole) without glazing, found behind the wall of the Sand Chamber. **C.** Base of a henkel pot with both sides glazed and reused as an oil lamp, which was found in the niche behind the small Millionary stalagmite. **D.** Henkel halfshell ceramic oil lamp with inner side glazing (Lamp 3) from the middle part of the Millionary Hall (first large collapse block). **E.** Henkel halfshell ceramic oil lamp with inner side glazing (Lamp 2) from the Sand Chamber (niche below central ceiling collapse block behind the wall). **F.** Pressed ceramic oil lamp (Lamp 1) from the Sand Chamber (niche below central ceiling collapse block behind the wall) (positions of finds in the cave see Fig. **5B**) (coll. Rabenstein Castle Museum).

Smashed Beer Glass Bottles of the brand "Held-Bräu" of the 50-70's comes from a local brewery in the next village.

Finally, **Ceramic Water Bottle** fragments were found again behind the Sand Chamber wall. These remains are from times after the 1950's.

Writing Tools and Militaria of the Bavarian King Ludwigs's I Time

Two half slate pencils and a shist plate fragment from the Reindeer Hall (center and lower area, Figs. **5B**, **7A - B**) seem to have been lost during early historical documentation and writings within the cave. A military helmet with a Golden oak leaf decoration with "L" (Fig. **7C**) was possibly deposited behind the Sand Chamber wall, which was comparable to military dresses of the King Ludwig I (1825-1848) regentship times from a "Bavarian helmet"[61]. Other militaria include the record of a pistol bullet cartridge, with oak acorn emblem (Fig. **7D**).

In the same corner close to the cave wall behind the artifical wall of the Sand Chamber, some larger cave bear and wolf bones, such as Jurassic fossils (ammonite and belemnite), must have been deposited in modern historic times by a former cave guide.

Figure 7. Modern historic times finds from the Sophie's Cave. **A.** Half slate pencil (initials F.G.) from the area of the lower reindeer hall. **B.** Half slate pencil from the middle Reindeer Hall. **C.** Golden oak leaf decoration from a Bavarian helmet with letter "L" of the regent times of King Ludwig I (1825-1848) from the Sand Chamber (behind the wall). **D.** Pistol bullet cartridge with oak emblem from the middle Reindeer Hall (coll. Rabenstein Castle Museum).

REFERENCES

[1]   Cuvier GLCFD de Baron. Sur les ossements du genre de l' ours, qui se trouvent en grande quantité dans certaines cavernes d`Allemagne et de Hongarie. Ann Mus Hist Nat 1806; 8: 3-25.

[2]   Heller F. Die berühmten Knochenhöhlen des fränkischen Jura und das Schicksal ihres Fundinhaltes. Nach zeitgenösischen Berichten und Quellen. Ber Naturwiss Ges Bayreuth 1966; 12: 5-20.

[3]   Rabeder G, Nagel D, Pacher M. Der Höhlenbär. Thorbecke Species 4. Stuttgart, Thorbecke 2000; p. 111.

[4]   Rabeder G. Neues vom Höhlenbären: Zur Morphologie der Backenzähne. Die Höhle 1983; 34(2): 67-85.

[5]   Rabeder G. Die Evolution des Höhlenbärgebisses. Mitt Kommiss Quartärf Österr Akad Wissensch 1999; 11: 1-102.

[6]   Diedrich C. Die oberpleistozäne Population von *Ursus spelaeus* Rosenmüller 1794 aus dem eiszeitlichen Fleckenhyänenhorst Perick-Höhlen von Hemer (Sauerland, NW Deutschland). Philippia 2006; 12(4): 275-346.

[7]   Diedrich C. Ice Age geomorphological Ahorn Valley and Ailsbach River terrace evolution– and its importance for the cave use possibilities by cave bears, top predators (hyenas, wolves and lions) and humans (Late Magdalénians) in the Frankonia Karst – case studies in the Sophie's Cave near Kirchahorn, Bavaria. Quat Sci J 2013; 62(2): 162-74.

[8]   Kaulich B, Rosendahl W. The Neonate cave bear skeleton from the Petershöhle near Velden (Franconia Alb, Germany). Abh Karst- Höhlenk 2002; 34: 12-6.

[9]   Rathgeber T. Die quartären Säugetier-Faunen der Bären- und Karlshöhle bei Erpfingen im Überblick. Laichinger Höhlenfreund 2003; 38(2): 107-44.

[10] Rosendahl W, Darga R. (Eds.) The Neue "Laubenstein-Bärenhöhle", Chiemgau/Bavarian Alps – the first alpine cave bear cave in Germany. Atti Mus Civ Stor Nat Trieste 2003; 49: 93-9.

[11] Groiss JT. Paläontologische Untersuchungen in der Zoolithenhöhle bei Burggeilenreuth. Ein vorläufiger Bericht Erlanger Forsch B Naturwissensch 1979; 5: 79-93.

[12] Techner K, Geyer M. Forschungsergebnisse aus dem Geisloch bei Oberfellendorf und benachbarten Höhlen um Muggendorf und Streitberg (Nördliche Frankenalb). Karst Höhle 1980; pp. 1-74.

[13] Diedrich C. Impact of the German Harz Mountain Weichselian ice-shield and valley glacier development onto Palaeolithics and megafauna disappearance. Quat Sci Rev 2013; 82: 167-98.

[14] Diedrich C. Evolution, Horste, Taphonomie und Prädatoren der Rübeländer Höhlenbären, Harz (Norddeutschland). Mitt Verb dt Höhlen- Karstf 2013; 59(1): 4-29.

[15] Münzel SC, Stiller M, Hofreiter M, Mittnik A, Conard NJ, Bocherens H. Pleistocene bears in the Swabian Jura (Germany): Genetic replacement, ecological displacement, extinctions and survival. Quat Int 2011; 245: 225-37.

[16] Hofreiter M. Genetic stability and replacement in late Pleistocene cave bear populations. Abh Karst-Höhlenk 2002; 34: 64-7.

[17] Rabeder G, Hofreiter M. Der neue Stammbaum der Höhlenbären. Die Höhle 2004; 55(1-4): 1-19.

[18] Stiller M, Baryshnikov G, Bocherens H, *et al.* Withering away - 25,000 years of genetic decline preceded cave bear extinction. Molec Biol Evol 2010; 27(5): 975-8.

[19] Rabeder G, Hofreiter M, Nagel D, Paabo S, Withalm G. Die neue Taxonomie der Höhlenbären. Abh Karst- Höhlenk 2002; 34: 68-9.

[20] Rabeder G, Hofreiter M, Nagel D, Whithalm G. New taxa of Alpine cave bears (Ursidae, Carnivora). Cah sci Dép Rhône-Mus Lyon 2004; 2: 49-67.

[21] Hilpert B. Studies of the morphology of the bears from the Steinberg-Höhlenruine near Hunas. Abh Naturhist Ges Nürnberg 2006; 45: 117-24.

[22] Hilpert B, Kaulich B. Eiszeitliche Bären aus der Frankenalb - Neue Ergebnisse zu den Höhlenbären aus dem Osterloch in Hegendorf, der Petershöhle bei Velden und der Gentnerhöhle bei Weidlwang. Mitt Verb dt Höhlen- Karstf 2006; 52(4): 106-13.

[23] Gümbel CW von. Geognostische Beschreibung des Königreichs Bayern. Band 4: Geognostische Beschreibung der Fränkischen Alb (Frankenjura) mit dem anstossenden Fränkischen Keupergebiete. Cassel, Fischer 1891; p. 761.

[24] Heller J. Muggendorf und seine Umgebungen oder die fränkische Schweiz. Ein Handbuch für Wanderer in diese Gegend, mit den Reiserouten und nothwendigen Notizen für Reisende. Neue, sehr vermehrte Auflage. Mit einer Karte. Bamberg, Dresch 1842; p. 234.

[25] Neischl A. Die Höhlen der fränkischen Schweiz und ihre Bedeutung für die Entstehung der dortigen Täler. Nürnberg, Schrag 1904; p. 95.

[26] Brückner K. Führer durch die fränkische und Hersbrücker Schweiz. 2 Auflage. Wunsiedel, Kohler 1907; p. 296.

[27] Habbe K-A. Der Karst der Fränkischen Alb - Formen, Prozesse, Datierungsprobleme. Die Fränkische Alb. Schr Zentr Inst fränk Landesk Univ Erlangen 1989; 28: 35-69.

[28] Meyer RKF, Schmidt-Kaler H. Wanderungen in die Erdgeschichte (5). Durch die fränkische Schweiz. München, Pfeil-Verlag 1992; p. 167.

[29] Kaulich B, Schaaf H. Kleiner Führer zu den Höhlen um Muggendorf. Nürnberg 1993; p. 125.

[30] Baier A. Hydrogeologie Frankens: Heilwässer, Wasserstollen und Karstquellen. Jb mittelrh geol Ver N F 2003; 85: 95-167.

[31] Heller J. Muggendorf und seine Umgebungen oder die fränkische Schweiz. Ein Handbuch für Wanderer in diese Gegend, mit den Reiserouten und nothwendigen Notizen für Reisende. Bamberg, Dresch 1829; p. 215.

[32] Cave cadastre FHKF (unpublished documents).

[33] Sieghardt A. Im Bannkreis der Wiesent. Kultur, Geschichts- und Landschaftsbilder aus der Fränkischen Schweiz. Band 1 Nürnberg. Koch 1925; p. 203.

[34] Kaulich B. Die Sophienhöhle bei Rabenstein. In: Voit G, Kaulich B, Rüfer W. (Eds.) Vom Land im Gebirg zur Fränkischen Schweiz. Eine Landschaft wird entdeckt. Erlangen: Palmen und Enke 1992; pp. 175-288.

[35] Goldfuss GA. Die Umgebungen von Muggendorf. Ein Taschenbuch für Freunde der Natur und Alterthumskunde. Erlangen 1810; p. 351.

[36] Esper JF. Ausführliche Nachricht von neuentdeckten Zoolithen unbekannter vierfüssiger Thiere und denen sie enthaltenden, so wie verschiedenen andern, denkwürdigen Grüften der Obergebürgischen Lande des Marggrafthums Bayreuth. Nürnberg 1774; p. 148.

[37] Buckland W. Reliquiae Diluvianae, or observations on the organic remains contained in caves, fissures, and diluvial gravel, and other geological phenomena, attesting the action of an universal deluge. London, J. Murray 1823; p. 303.

[38] Cramer H. Bärenschliffe in fränkischen Höhlen. Fränkische Heimat 1931; 10(7): 206-9.

[39] Leja, E. Bärenschlafplätze in einer Höhle der Frankenalb (Bayern). Die Höhle 1999; 2: 65-70.

[40] Rosendahl S, Döppes D. Trace fossils from bears in caves of Germany and Austria. Sci Ann School Geol Arist Univ Thessaloniki 2006; spec vol 98: 241-9.

[41] Diedrich C. Ichnological and ethological studies in one of Europe's famous bear den in the Urşilor Cave (Carpathians, Romania). Ichnos 2011; 18(1): 9-26.

[42] Diedrich C. Holotype skulls, stratigraphy, bone taphonomy and excavation history in the Zoolithen Cave and new theory about Esper's "great deluge". Quat Sci J 2014; 63(1): 78-98.

[43] Rosenmüller JC. Die Merkwürdigkeiten der Gegend um Muggendorf. Unger 1804; p. 90.

[44] Diedrich C. The largest European lion *Panthera leo spelaea* (Goldfuss) population from the Zoolithen Cave, Germany – specialized cave bear predators of Europe. Hist Biol 2010; 23(2-3): 271-311.

[45] Diedrich C. Late Ice Age wolves as cave bear scavengers in the Sophie's Cave of Germany – extinctions of cave bears as result of climate/habitat change and large carnivore predation stress in Europe. ISRN Zoology 2013; 1-25.

[46] Diedrich C. The Late Pleistocene spotted hyena *Crocuta crocuta spelaea* (Goldfuss 1823) population with its type specimens from the Zoolithen Cave at Gaillenreuth (Bavaria, South Germany) – a hyena cub raising den of specialized cave bear scavengers in boreal forest environments of Central Europe. Hist Biol 2011; pp. 1-33.

[47] Diedrich C. The rediscovered cave bear "*Ursus spelaeus* Rosenmüller 1794" holotype of the Zoolithen Cave (Germany) from the historic Rosenmüller collection. Acta Carsol Slov 2009; 47(1): 25-32.

[48] Schabdach H. Die Sophie's Cave im Ailsbachtal. Wunderwelt unter Tage. Ebermannstadt, Verlag Reinhold Lippert 1998; p. 47.

[49] Wagner R. Ueber die neu entdeckte Zoolithenhöhle bey Rabenstein. Bayer Ann 1833; 47: 313-15.

[50] Holle JW. Die neu entdeckte Kochshöhle oder die Höhlenkönigin im königlichen Landgerichte Hollfeld-Waischenfeld. Bayer Ann 1833; 26: 197-8.

[51] Sternberg K. Vortrag des Präsidenten Grafen Kaspar Sternberg in der allgemeinen Versammlung des böhmischen Museums in Prag. Verh Ges vaterl Mus Böhmen Prag 1835; pp. 12-30.

[52] Staatsarchiv Bayreuth, unpublished documents K3FVVIII/321. Regierung des Obermainkreises/Oberfranken 1833.

[53] Staatsarchiv Bayreuth, unpublished documents K3FVVIII/321. Regierung des Obermainkreises/Oberfranken 1837-44.

[54] Staatsarchiv Bayreuth, unpublished documents K17/8868. Landgericht Hollfeld 1932.

[55] Steguweit L. Untersuchungen der Schichtenfolge am Vorplatz der Sophie's Cave bei Burg Rabenstein. In: Greipl EJ, Sommer S, Päffgen B. (Eds.) Das archäologische Jahr in Bayern 2008. Theiss-Verlag 2008; pp. 11-3.

[56] Bauer I, Endres W, Kerkhoff-Hader B, Koch R, Stephan H-G. Leitfaden zur Keramikbeschreibung (Mittelalter-Neuzeit). Terminologie-Typologie-Technologie. Kat Prähist Staatsslg Beih 2005; 2: 1-200.

[57] Gross U. Schwäbisch Gmünd-Brandstätt: Keramikfunde aus einer Kellerverfüllung der Zeit um 1800. Eine vorläufige Übersicht. Teil 1: Irdenware. Fundb Baden-Württembeerg 1999; 23: 667-720.

[58] Mattes J. Reisen ins Unterirdische. Vienna 2013; p. 489.

[59] Shaw T. Aspects of the history of Slovene Karst 1545-2008. Ljubljana 2010; p. 306.

[60] Bernard C. Die Gefäßkeramik saarländischer Burgen - ein Forschungsdesiderat: Erste Einblicke. In: H-J Kühn (Ed.) Beiträge zum 1. Saarländischen Burgensymposion am 31. März 2007 in Saarbrücken. Saarbrücken/Münster 2009; pp. 11-46.

[61] Funcken L, Funcken F. Historische Uniformen 18. und 19 Jahrhundert. München: Orbis Verlag 2000; p. 480.

GEOLOGY OF THE CAVE ROCKS IN UPPER FRANCONIA

Abstract: The white-yellowish, massive Upper Jurassic dolomite reef rocks (155-150 My) of Upper Franconia are famous for climbers, because of its rich cavities, partly caused initially by burrowing marine crustaceans (= Lochkalke), partly due to rock-weathering and cave erosion. It is one of the most cave-rich regions in Europe counting several hundreds of mainly smaller caves, and few very large cave systems. Those are situated on the Upper Franconia Plateau, which is cut by the Wiesent and smaller branching river valleys. The Late Jurassic fossils (ammonite steinkerns, and metasomatic changed silified and originally calcite reef fossils) found within Pleistocene sediments of the Sophie's Cave supports reconstructing the Pliocene plateau and Pleistocene valley erosion history and geomorphological changes in the surroundings.

Keywords: Upper Franconia dolomite karst, European cave-rich region, Late (White) Jurassic reefs, metasomatic changed reef fossils, geomorphology change reconstruction.

GENERAL GEOLOGY OF CAVE-RICH UPPER FRANCONIA

Upper Franconia has two different main geological regions: the western Mesozoic more or less horizontally deposited marine series ranging from Jurassic to Cretaceous, and the eastern to Czech Republic overlapping old Bohemian Massif rocks consisting of Palaeozoic sediments, but mainly of metamorphites, plutonites and vulcanites (Fig. **1**) [1 - 3]. The cave-rich-region is limited on the western part within Late Juarassic marine sediments. In most areas of Upper Franconia the morphological slope (= "Albtrauf" in German) [1 - 3] is built of those cave-rich massive Late Jurassic dolomites and thick-layered limestones [4], which were built within the 150 My old Late Jurassic (in southern Germany, = White Jurassic, Fig. **1**) [1 - 4] under a warm, tropical, shallow shelf and lagoon in Germany (Fig. **2**) [4].

Cajus G. Diedrich

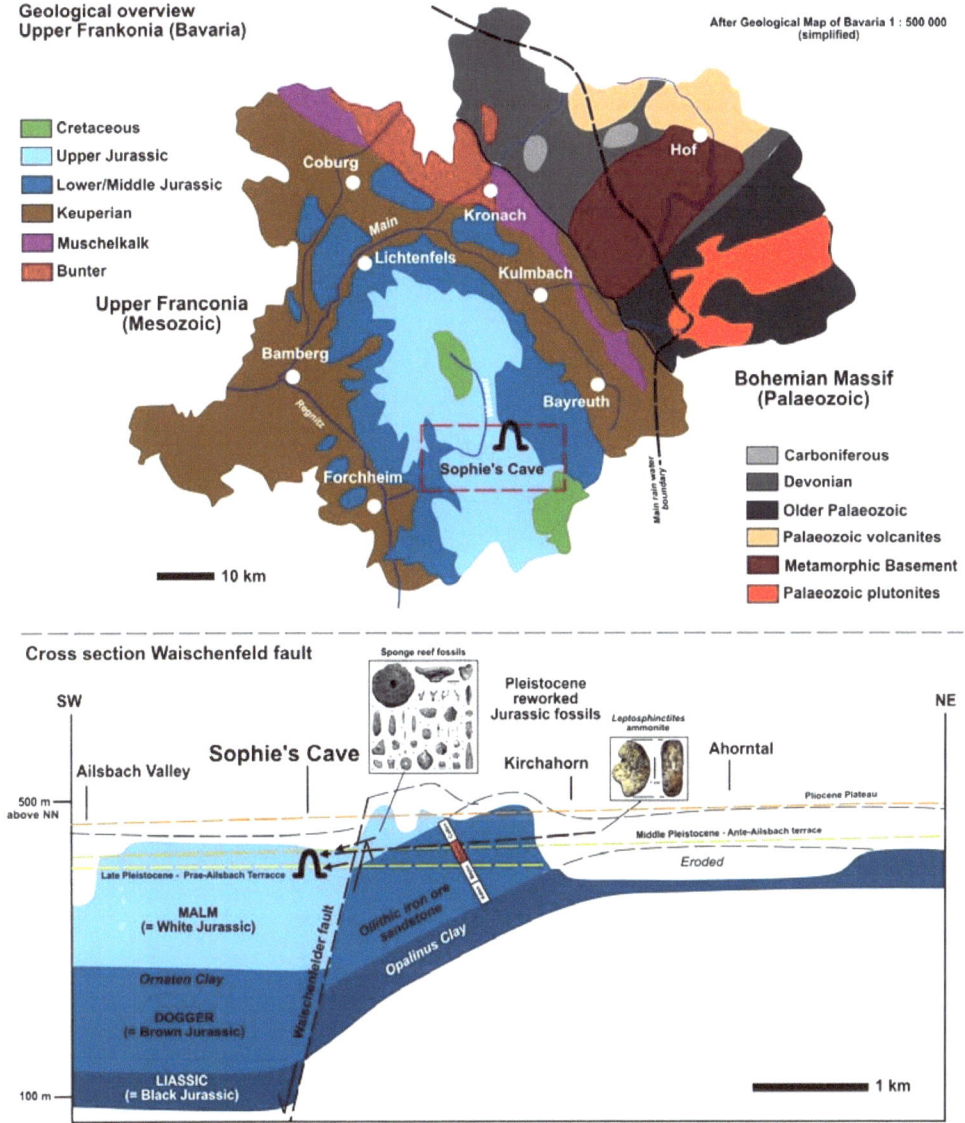

Figure 1. Geological overview of Upper Franconia and cross-section along the Ailsbach Valley, Ahorn Valley depression and Waischenfeld fault, and former elevations of the river terraces. The fossils found in the Middle/Late Pleistocene sediments of the Sophie's Cave are from the surrounding Brown and White Jurassic, whereas quartz pebbles are from nearby Lower Cretaceous sandstones (geology modified after [1, 2]).

The up to several decameters massive dolomites, well exposed above the Sophie's

Cave entrance (Fig. **2**), were deposited on the western margin of the Bohemian Massif. This Massif built a main land area during the Central European Jurassic Sea time, whereas during this same time period along the northern German and along the London-Brabant Massif mainland coasts large dinosaurs migrated [5].

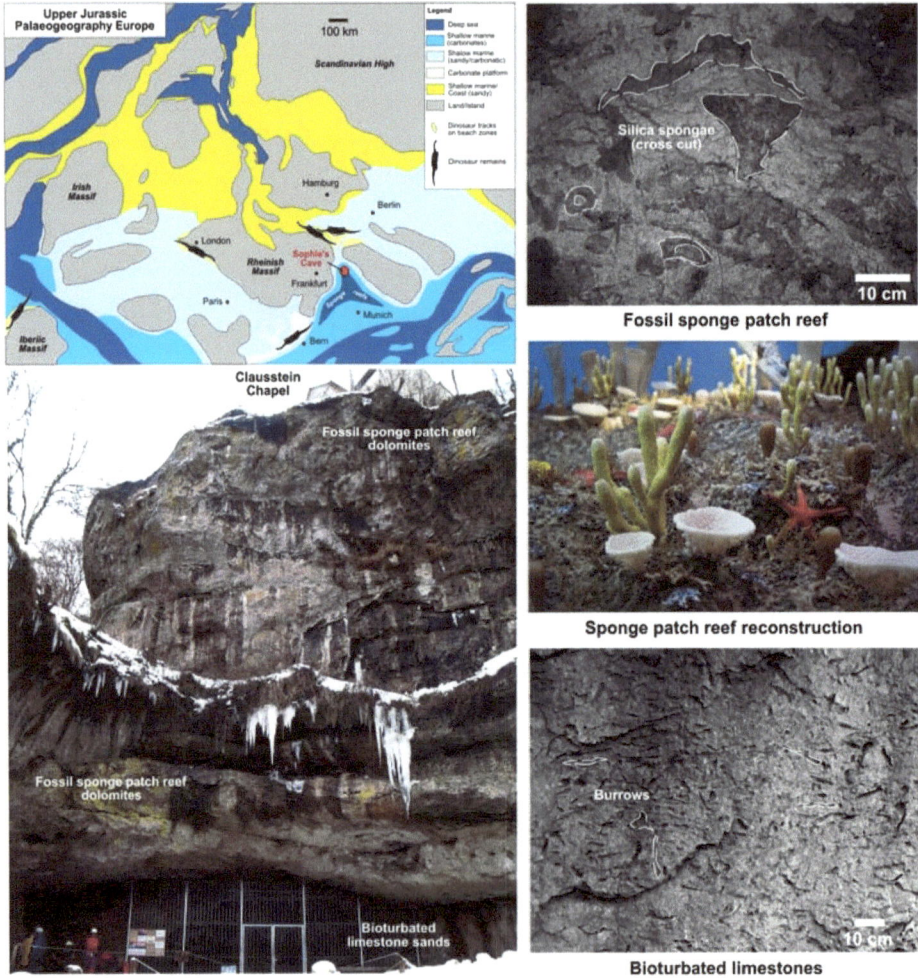

Figure 2. Upper Jurassic Palaeogeography, dinosaur sites and Bavarian sponge reef and lagoon facies within the northern Tethys Ocean [5]. Shrimp (such as similar to modern mud shrimp) bioturbated (= *Thalassinoides/Spongeliomorpha* burrows) in shallow marine Upper Jurassic calcareous sediments (= "Lochkalke") between the reef complexes (wall entrance Sophie's Cave), and in the cave relief-like weathered silificated spongae in the Collapse Hall, such as reconstruction of small sponge-patch reef builders (reef reconstruction, Staatliches Museum Naturkunde Stuttgart).

Between those reef complexes the famous central Bavarian Plattenkalke with its "early bird" *Archaeopteryx* finds formed, which contain rare small dinosaurs or lizards, marine crocodiles and turtles, but mainly marine fish, sharks and invertebrates. Those platy limestones also contain land plants in a high biodiversity which accumulated under lagoonary situations. Those plants originate from the mainland of a tropical dinosaur period [6]. In more northern Bavaria (Upper Franconia) between the reefs up to decameter wide-sized calassianid crustacean bioturbated dolomitic carbonate sands were deposited in shallow sea conditions. Today those softer rocks built morphological depressions within the landscape and built in some areas rock shelters [1, 4]. Those are in change with convex relief and stronger cemented harder sponge patchreef structures (Fig. **2**) [1].

The White-Jura Dolomite Reef-Lagoon Rocks (155-150 My)

The dome-shaped halls in the Sophie's Cave, which are connected by narrow tunnel-like passages, have their ceilings along the subhorizontal, horizontal and wavy bedding planes of the two main facies or rock types: those are the massive sponge reef dolomites (Fig. **2**), and the crustacean burrowed (*Thalassinoides/ Spongeliomorpha* ichnites) "hole chalks" (= " Lochkalke " in German; Fig. **2**) [1]. In the transition between the decameter thick scaled patch reefs, the "Lochkalke" are intercalated filling the former depressions between the dom-shaped reef complexes [1].

The change of those two main facies and different weathering (reef – convex mor-phology, and more robust rocks; bioturbated dolomitic limestones – concave and more quick weathering) [1] explains also the changing relief of the 30 m high rock wall above the Sophie's Cave entrance (Fig. **2**).

Originally, the rocks were shallow water carbonates that cemented diagenetically to consolidated limestone mud [7]. Late diagenetically, starting within the Alpine orogeneses and uplifting of the Upper Franconia Mountain region, the chemistry of the carbonate rocks changed by a dolomitization process [7]. Magnesium-rich (Mg) pore waters (phreatic) replaced the calcium (Ca), which caused a change from a calcitic limestone ($CaCo_3$) to a dolomite ($CaMgCo_3$) [7].

The Upper Jurassic sponge reef structures are only 1-3 meters in height and up to 10 meters in width, which can be easily seen in the Entrance Hall, or the Collapse Hall of the Sophie's Cave (Fig. **2**). There was no large reef complex in the northern Bavarian Late Jurassic lagoons, such as what is seen in modern coral reef build up [4, 8]. Instead, there were isolated patch reefs, which formed along the Upper Jurassic lagoon in water depths of 3-15 m in the northern Bavarian Lagoon region [4, 6, 8]. The boundaries between patch reef and intrareef refill are originally carbonate mud layers which depressions [4, 8] can be seen between the dome shapes on the Sophie's Cave ceilings (Ahorn, Clausstein, Collapse halls).

The Sponge Patch Reef Fossils from Sophie's Cave

In the Sophie's Cave, the fossils from the dolomites weathered naturally, starting in the Pliocene when the cave formed [9]. This chemically perfect preparation by natural carbonic acid of the silificated invertebrate fossils continued in the Pleistocene [9]. Those fossils were enriched with well preserved specimens in the Late Pleistocene gravels and sands, in which they built a high amount of the matrix [9]. The best specimens who were sieved and are figured herein (Fig. **3**) come from the sands and gravels of the Clausstein Hall, which represent the main common benthic reef organism types which are found all over the northern Bavarian Late Jurassic reef facies [4, 8]. This small invertebrate sponge reef fauna is already well listed, with more than 300 species from other localities [4, 8].

By calcite or aragonite carbonate built up reef adapted organism skeletons of the Late Jurassic were metasomatic replaced during the dolomitization process (also Sophie's Cave material), which detail structures were often lost. The replaced mineral was silica (SiO_2) from acids which originated from spongae sklerocytes [1, 7, 8]. The spongae were also silicified [8], and those also built generally the main part and centre of silica concreations (= "Hornstein", in German) [8]. Those silica concretions and sponge fragments, and the silicified small Late Jurassic reef fossils [8] even survived further Pleistocene weathering and water transport, and were enriched in the Middle to Late Pleistocene sands/gravels of the Sophie's Cave. Those must originate mainly of the weathering surrounding cave walls or nearby because those are not enrolled or polished and have generally perfect details preserved (Fig. **3**). The fossils of a small reef building organism biodiversity can be reconstructed partly from the Sophie's Cave material that was sieved mainly from the Claustein Hall Late Pleistocene sands/gravels (Fig. **3**). Besides the smaller macrofauna, bacteria and blue-green algae were responsible for the reef built up, by chemical cementation and binding of carbonates, which often cover complete spongae colonies [4, 8]. Up to two meters long crinoids settled in colonies in the clean, warm, and oxygen-rich hypersaline lagoon water within the patch reefs, but fell apart, into thousands of pieces, similar as the corona and spines of different reef adapted sea urchins [4, 8] (Fig. **3**).

The main sponge forms can be studied better in their cross-sections which are found all over in the cave walls, with best sources in the Collapse Hall (Fig. **2**). On the down-side of the spongae, and between their colonies, small and fan-branched corals, cemented brachiopods, bivalves such as oysters, and finally, bryozoan colonies were the reef builders [4, 8]. The spongae are of two groups, the "silica" and "calcareous spongae" [4]. The calcareous forms are very different, with small, elongated, tube-like and branching forms, looking like bryozoan fan-like colonies to which *Neuropora* sp. (Figs. **3.4-5**) belongs. The most abundant spongae are the silica spongae, such as tube-like *Tremadictyon reticulatum* (Fig. **3.1**) the platy form (Fig. **3.2**), and conical honeycomb-like morphology type *Sporadopyle obliqua* (Fig. **3.3**). Corals are represented by fragments only of the fan-like type *Goniocora* sp. and *Enallhelia elegans* (Fig. **3.6-7**), whereas bryozoan colonies are represented by triaxonic forms of unknown species (Fig. **3.8**) [4, 8]. Serpulid worms have left their tubes of the species *Glomerula gordialis,* which built mostly convoluted forms which are attached and cemented on spongae or shells of other marine organisms (Fig. **3.9**) [4, 8]. The small oyster *Deltoideum delta* and zigzag-oyster *Arctostrea gregaria* (Fig. **3.10**) were also cemented on other organisms, or even on the hard grounds of the reef [4, 8]. Pectinid vagile clams were in those reefs *Radulopecten* sp. (Fig. **3.11**) [4, 8].

Very common and best visible are the coral-reef adapted thickened non-spiky spined sea urchin and fewer the corona fragments of *Plegiocidaris* sp. (Fig. **3.12-15**) or *Merocidaris* sp. (Fig. **3.16**), which species hid in the niches between the reefs [4, 8]. Besides those, needle-like long spines of other small-sized benthic reef sea urchins of unclear genera are common (Fig. **3.21-23**). Conspicous are the spiky long and distally flattened spines of the largest reef echinid *Rhabdocidaris*

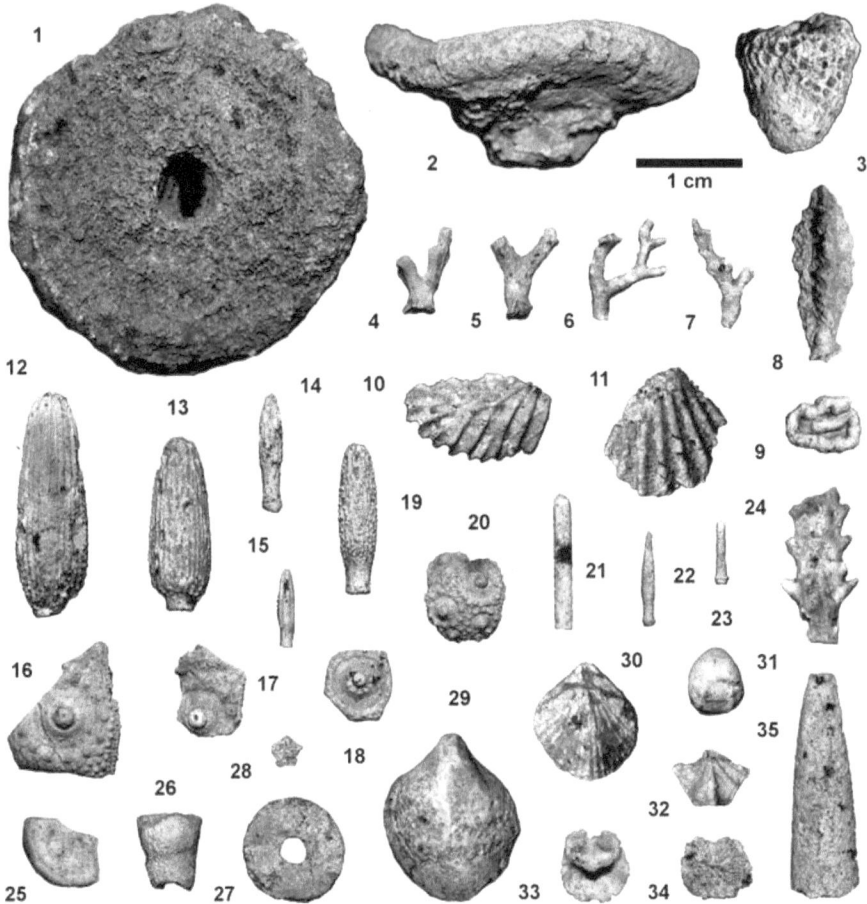

Figure 3. From Late Pleistocene river terrace sands/gravels of the Clausstein Hall (section D) sieved silificated Upper Jurassic reef fossils. **1.** Silica sponge *Tremadictyon reticulatum* in cross-section. **2.** Silica sponge *Sporadopyle obliqua* in Seitenansicht. **3.** Silica sponge *Sporadopyle obliqua.* **4-5.** Calcareous sponge *Neuropora spinosa* colony fragments. **6.** Fan-like coral *Goniocora* sp. fragment. **7.** Fan-like coral *Enallhelia elegans* fragment. **8.** Bryozoan indet. fragment. **9.** Tube worm *Glomerula gordialis.* **10.** Zigzag oyster *Arctostrea gragaria.* **11.** Pectinid bivalve *Radulopecten* sp. fragment. **12-15.** Spines of the reef-adapted sea urchin *Plegiocidaris* sp. **16-18.** Sea urchin *Plagiocidaris/ Rhabdocidaris* corona plates. **19.** Spines of the reef-adapted sea urchin *Merocidaris* sp. **20.** Corona part of the sea urchin ?*Merocidaris* sp. **21-23.** Spines of small reef sea urchins. **24.** Sea urchin *Rhabdocidaris* sp. spine fragment **25.** Sea star *Tylasteria* sp. arm plate. **26-27.***Millericrinus* crinoid remains (crown and stem elements). **28.***Balanocrinus* sp. crinoid stem element. **29.** Most abundant brachiopod *Juralina* sp. steinkern. **30.** Brachiopod *Dictyothyropsis loricata.* **31.** Brachiopod *Monticlarella* sp. **32.** Brachiopod *Ismenia recta.* **33.** Brachiopod *Craniscus tripartitus* cemented valve. **34.** Brachiopod *Craniscus tripartitus* free valve. **35.** Belemnite *Hibolithes semisulcatus* rostrum fragment (coll. Rabenstein Castle Museum).

sp. (Fig. **3.24**) [4, 8]. Seastars are represented by single plates of *Tylasteria* sp. (Fig. **3.25**); crinoids are preserved by round mill-stone-like stem pieces and crown elements of *Millericrinus* sp. (Fig. **3.26-27**), but also by another form *Balanocrinus* sp., that has star-like stem elements (Fig. **3.28**) [4, 8]. Several brachiopods of 3-30 mm in size are represented mostly by the 30 mm large terebratulid *Juralina* sp. (Fig. **3.29**), which can often be found sticking out of the cave walls even in typical "colonies" [4, 8]. Other small brachiopods are *Ismenia recta* (Fig. **3.32**), *Monticlarella* sp. (Fig. **3.31**), *Dictyothyropsis loricata* (Fig. **3.30**) or on spongae attached *Craniscus tripartitus* (Fig. **3.33-34**) in the patch reef complexes [4, 8]. The only cephalopod remains (no ammonite records yet from Sophie's Cave dolomites) are belemnite rostra of *e.g. Hibolithes semisulcatus* (Fig. **3.35**), but also those original calcite rostra are silified now.

REFERENCES

[1] Habbe K-A. Der Karst der Fränkischen Alb - Formen, Prozesse, Datierungsprobleme. Die Fränkische Alb. Schr Zentr Inst fränk Landesk Univ Erlangen 1989; 28: 35-69.

[2] Meyer RKF, Schmidt-Kaler H. Wanderungen in die Erdgeschichte (5). Durch die fränkische Schweiz. München, Pfeil-Verlag 1992; p. 167.

[3] Freyberg von B. Tektonische Karte der Fränkischen Alb und ihrer Umgebung. Erlanger Geol Abh 1969; 77: 1-81.

[4] Lauxmann U, Schweigert G, Kapitzke M. Die Schwamm- und Korallenriffe der Schwäbischen Alb. Erdgesch mitteleurop Reg 1998; 2: 117-28.

[5] Diedrich C. New dinosaur tracks and dinosaur remains of Northwest Germany - palichnostratigraphy and the megatracksite concept in the Kimmeridgian (Upper Jurassic). Palaeobiodiv Palaeoenvironm 2010; 91(2): 129-55.

[6] Mäuser M. Frankenland am Jurastrand. Versteinerte Schätze aus der Wattendorfer Lagune. Friedrich Pfeil-verlag 2008; p. 60.

[7] Burger D. Dolomite weathering and micromorphology of paleosoils in the Franconian Jura. Catena Supplement 1989; 15: 261-7.

[8] Sauerborn U. Die Korallenkalk-Fauna von Nattheim. In: Weidert WK. (Ed.). Klassische Fundstellen der Paläontologie 1. Korb 1988; pp. 77-84.

[9] Diedrich C. Ice Age geomorphological Ahorn Valley and Ailsbach River terrace evolution– and its importance for the cave use possibilities by cave bears, top predators (hyenas, wolves and lions) and humans (Late Magdalénians) in the Frankonia Karst – case studies in the Sophie's Cave near Kirchahorn, Bavaria. Quat Sci J 2013; 62(2): 162-74.

CHAPTER 3

PLIO- TO MIDDLE PLEISTOCENE SEDIMENTOLOGY, CAVE GENESIS AND AILSBACH VALLEY GEOMORPHOLOGY

Abstract: The Sophie's Cave in southern Germany, in the middle high elevated (max 550 m a.s.l.) mountains of northern Bavaria, was formed by Pliocene subsurface ground waters of the Upper Franconia Jurassic Plateau (about 440 a.s.l.). In the ponor cave stage of Early Pliocene age, the horizontal system which started to refill partly with about 3-4 m iron-and manganese-rich clays and dolomite ash sands (= coloured series). Within the intermediate cave stage in the Early-Middle Pleistocene, the Ailsbach River valley lowered from 440 to 420 m a.s.l. In the Middle Pleistocene, fluvial sediment intruded only from the valley side into the Sophie's Cave from above the Clausstein Hall vertical shaft consisting of 8 m thick river terrace clay, sand and gravel (= "yellow series"). A first Middle Pleistocene (?Holsteinian Interglacial) spel-eothem generation formed on the top. Middle Pleistocene marten *Martes* sp. used some parts at minimum in the Clauststein Hall as a den and left some tracks on muds being the first known Middle Pleistocene footprints named herein *Martichnus desseri* nov. ichnogen. and ichnosp. in Europe, which were casted and preserved by the speleothem layer. These Middle Pleistocene cave sediment and speleothems eroded somehow within the late Middle Pleistocene (?Saalian) in the valley sided cave branches by intruding floods.

Keywords: Pliocene to Middle Pleistocene, sedimentology, Ailsbach Valley geomorphology, *Martichnus desseri* nov. ichnog. and ichnosp., marten den.

PLIOCENE/EARLY PLEISTOCENE (5.3-1.8 MY) – CAVE GENESIS/ REFILL

The Ahorn Valley has the most dense number of caves in the Upper Franconia karst, although most of them are only small clefts or cavities [1 - 11]. The really large caves of the region, which are mostly horizontal ponor caves in their initial stages, are the König Ludwigs Cave, with its large entrance portal and one single, giant chamber (385 m a.s.l., historically also named "Kühloch" [12]). This cave is

Cajus G. Diedrich

situated within the Ahorn Valley opposite the higher elevated Sophie's Cave (411 m a.s.l.). Other larger caves at different elevations between 460 to 420 m a.s.l. with Late Middle to Late Pleistocene vertebrate bone content and bonebeds are the Große Teufels Cave, Moggaster Cave, Zoolithen Cave, or Geisloch Cave such as Zahnloch Cave (see Chapter 1, Fig. **1**) [13, 14]. Smaller caves contain lesser amounts of non-studied or species determined or dated cave bear remains: Neideck Cave, Wunder Cave and Esper Cave [13].

The karstification of the dolomite reef rocks have a longer earth history record starting already in the Late Cretaceous and Palaeogene [15], which can be reconstructed in detail following classical cave genesis and filling and speleogenesis models [16 - 22] which can be transferred also to the Sophie's Cave, which will be a guide for many other similar elevated caves around 410 m a.s.l, especially of Upper Franconia (*e.g.* Große Teufels Cave). Using sediment grain sizes (clay, silt, sand, gravel), the types of sands (dolomite ash, silica sand), their different colouring (manganese - black, iron limonite – orange/red), and the presence and absence of Jurassic fossils (see Chapter 2, Fig. **3**) from different originating strata (Fig. **1**), the refill history of the cave with fluvial/(?glacial) sediments can be subdivided into three main phases [13].

The cave forming started within the groundwaters of the Pliocene plateau, along the more or less horizontal layer discordances, but also diagonally within the clefts (Fig. **1**). The Sophie's Cave eroded into the massive sponge reef dolomites of the Late Jurassic (Malm Delta), with massive rocks building the plateau and highest peaks within Upper Franconia (Chapter 2, Fig. **1 - 2**) [13 - 15]. The passages are oriented mainly on cleft systems running NNE-SSW [23]. The cave is near the Waischenfeld Fault, where the Mesozoic layers moved along a main fault (Chapter 2, Fig. **1**) [24]. The Ahorn Valley saddle structure [24] was eroded (calculated on the Sophie's Cave sediments) already by the Pliocene, but was further deepened in the Middle Pleistocene, and received its bowl-like basin structure in the Late Pleistocene (Chapter 2, Fig. **1**) [13]. The ancient Pliocene cave river (elevation about 410 m a.s.l.) cut deeper and deeper into the first pipe system and built most of the cave system known as ponor cave, similar to all the surrounding caves (*e.g.* Bing Cave [25]).

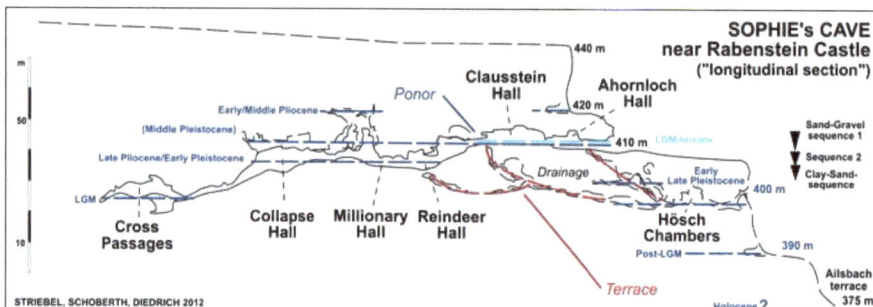

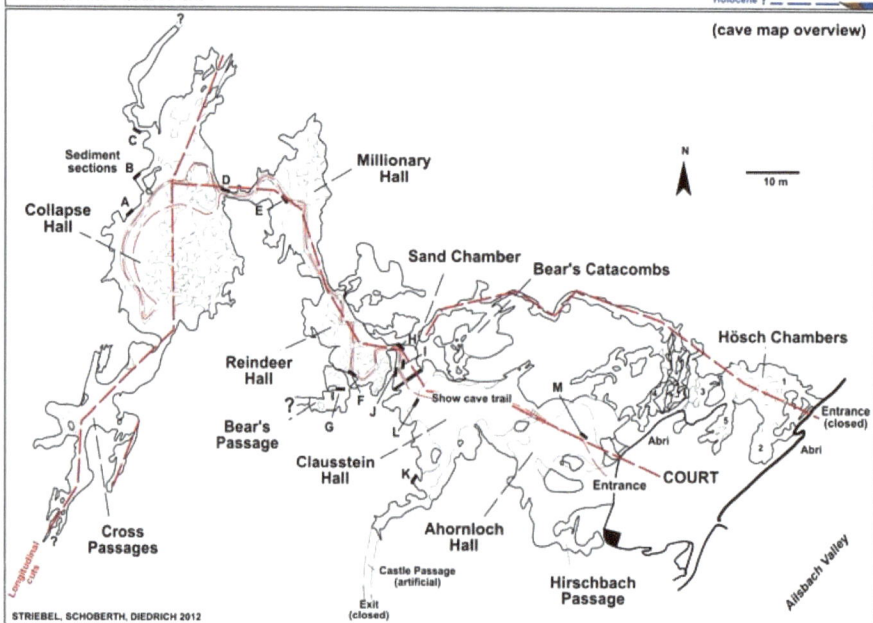

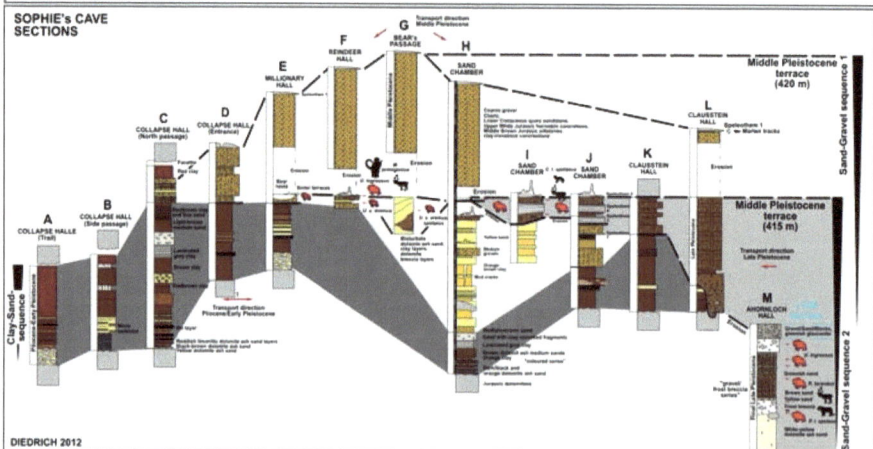

Figure 1. Sections in the Sophie's Cave, which are from three main and different aged cave genesis and river terrace sediment refill periods (see positions of sections above in the cave map, composed from [13]) (cave map outlines after unpublished map of Schobert an Schabdach, cross-section after unpublished map of Striebel).

The turbulent running water of the about 30 meters deep subsurface stream (Fig. **2**) left small-to-large scaled facets on the ceilings, typical in cross-cut key-hole passages (= typical pressure pipes [16]), or even typical large scour troughs [16] on the ceilings of the Ahornloch, Clausstein, and Collapse halls (Fig. **2**). Additionally, several smaller and often branched passages formed under high water pressure over the horizontal dolomite layer bedding planes and diagonal clefts [23]. The small facets on the ceilings of the Collapse Hall (Figs. **2 - 3**) indicate higher floating speeds [16 - 18] and are in cross-section elliptically asymmetrical. Due to their forms – in floating direction flat, and opposite direction steep margins [16] – the flow direction can be reconstructed in the cave to the (today's) entrance direction (also valley wards, Fig. **2**). The subsurface river must have come from the Collapse Hall and its Northern Passage (from the Plateau, Fig. **2A**) and was floating to the (today's) entrance. There might have been a spring on the Plateau, or farther on the plateau margin (Fig. **2A**).

The about one to four meters in sediment thickness initial fluvial refilled passages, chambers, and halls mainly consist of clay to fine sand grain-sized typical subsurface river deposits [20 - 22] ("coloured series", Fig. **3**) and were located in most places in the Sophie's Cave, in parts, or in sections at their bases [13]. Originally, the Sophie's Cave system was possibly angeled within the Pliocene of only about 0.5°. The age of the first relic "colored series" refill is not dated with faunal remains indirectly or absolute dating methods yet. It seems, after comparisons to Pliocene/Early Pleistocene micromammal assemblages from different even higher elevated caves (all over 460 m a.s.l.) [26 - 29] and first published coarse landscape genesis models of Upper Franconia [15], the oldest refill must have happened within the Pliocene to Early Pleistocene (5.3 to 1.8 My, or up to 1.4 My using the chronostratigraphy sensu [30]). The subsurface stream imported mainly clay minerals to which different metal ions attached, such as manganese (black) and iron (red-orange, Fig. **3**) being typical in many autochthonous cave sediments [20 - 22]. Those minerals originate from the continous dolomite erosion in the ponor cave itself, which also caused the production of typical dolomite ash sands that accumulated in several cm to dc thick layers, which built the famous historically mentioned "golden sands", consisting only of cube, and not well-rounded dolomite crystals [31 - 33]. During their short-distance water transport within the cave, they were additionally enriched with small fossil fragments which weathered from of the spongae reef dolomites (cf. Late Jurassic reef fossils in Chapter 2, Fig. 3).

Earth Quake Signs in Sediments

In some places microtectonic structures within the Plio-/Early Pleistocene sediments (coloured series) of the Sophie's Cave can be observed. The structures are obviously of tectonically shape origin compared to other similar aged Plio-/Early Pleistocene at nearby localities in northwest Bohemia [34] which are step- or horst-like structures in cm scale only, with best examples in the Sand Chamber (Fig. 4). Such microtectonic structures are similarly recorded from the ?Early/Middle Pleistocene (dolomite ash and red clay series) sediments of the Zoolithen Cave [14] in the Wiesent valley, or from caves in the Pegnitz Valley of Upper Franconia, which latter "earth quake" influenced sediment structures were also dated into the Middle Pleistocene [35].

Younger reports are from late Middle/Late Pleistocene gypsum karst fluvial sediments north of the Harz Mountains [36]. It seems, uplifting of the Upper Franconia Jurassic Plateau happened in the Plio-/Early Pleistocene continuously as result of the northwest Bohemian Massif "volcanic region" (see Chapter, Fig. 1) uplift [34], which still continues to this day. Small earth quakes in this Bohemian Massif region of northwest Czech Republic are known also from historic times [34]. The most recent newspaper reported earth quake (3.8 on skala) of the northwest Bohemian Massif on the 04.08.2014 was also recognizable in Upper Franconia [37].

Obviously, after the earthquake period and uplifting processes, the horizontal Sophie's Cave system was rotated within only/or since the Early Pleistocene to the valley side about 1-2° (compare Figs. 2 and 4, or Chapter 3, Fig. 1).

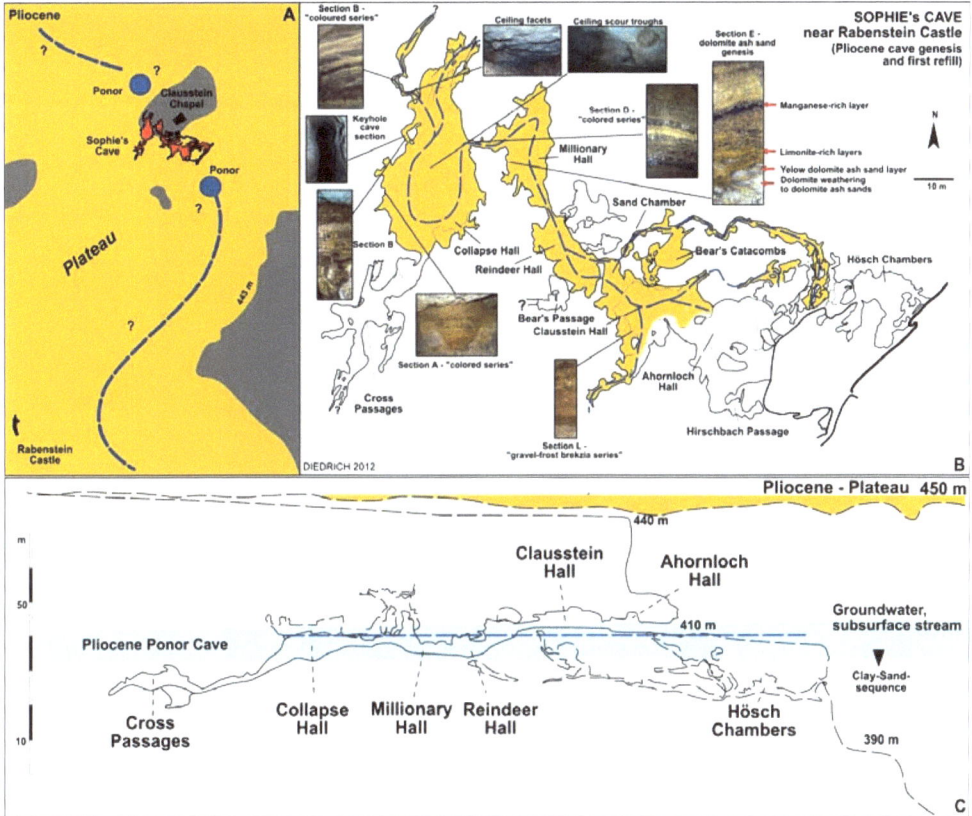

Figure 2. Cave genesis and first refill in the Pliocene/Early Pleistocene (5.3 to 1.8 My). **A.** Plateau reconstruction. **B.** Cave map of the Sophie's Cave in dorsal and lateral **C.** cross-cut (section outlines after T. Striebel) views, with filled areas, sections, and running water facets. Cave and aquifer (= ground water) of the ponor cave in cross section (back rotated about 1-3°) (modified after [13]).

LATE EARLY PLEISTOCENE (1.8-0.8 MY)? – FURTHER CAVE REFILL

Unclear remains the time frame after the ponor stage cave genesis, when the Ante-Alisbach River valley and terraces started to form, eroding between 450 and 420 m a.s.l. into the dolomite rocks. This can be estimated only on the steep cliffs above the Sophie's Cave entrance and along the valley in the upper elevations, because sediments of this period in the cave are missing [13]. The early valley genesis sediment records within the cave (Fig. **3**) can not yet be clearly correlated to other similar elevated caves yet, because of missing studies. Possibly, some of the "coloured series" falls within this time frame, too. It is certainly no early Early

Pleistocene river terrace fills that were deposited in the Sophie's Cave, because the terraces (being completely eroded outside the caves in the valleys) were still too high to reach open cavities or vertical shafts of the Sophie's Cave system (Fig. **2**).

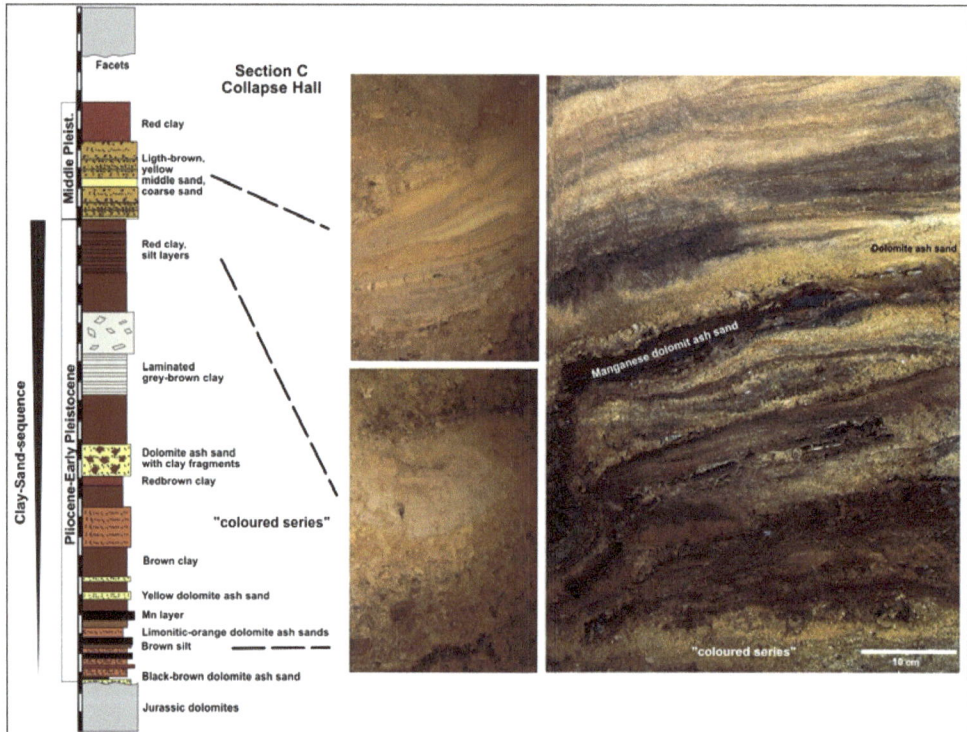

Figure 3. Key hole section (section C) refilled with the "colored series" of Pliocene/Early Pleistocene and finally also a few Middle Pleistocene ("yellow series") of the North Passage in the Collapse Hall, Sophie's Cave (modified after [13]).

MIDDLE PLEISTOCENE (780.000-200.000 BP) – ANTE-AILSBACH RIVER TERRACE

The Ante-Ailsbach River terrace was somehow within the Middle Pleistocene (final stage), about 420 m a.s.l., which is today the elevation of the first "dolomite plateau", 9 m above the Sophie's Cave entrance floor level. At this elevation, the terrace reached open "branches" of the upper parts of the Sophie's Cave, in which fluvial terrace sediments were washed only into some parts of the cave (Figs. **5** -

6) [13]. The infill origin and transport direction can be reconstructed very well. Those sediments intruded from above into the Clausstein Hall (about eight meter thick series) and Reindeer Hall, mainly from a today's blocked/refilled vertical shaft (Fig. **5**). The massive, up to eight meter thick sediment conus reached deeper parts of the cave, the Bear's Passage (about one meter thick series), the Sand Chamber (about 3.5 m thick series, Fig. **6**), and finally, with much lesser amounts and thickness, into the Millionary Hall (cf. section correlations in Fig. **1**) [13].

The lowermost Middle Pleistocene fluvial sediments of the "yellow series" consist of layers of cm to dcm-thick changing clay, silt, sand and medium-sized gravel (max. 1 cm). These are best documented and preserved after section preparations with a full fluvial flood sequence only in the Sand Chamber (Figs. **1, 5 - 6**), which was eroded in its upper parts much later (Fig. **6**) [13].

Compared to a complete river terrace flood sediment sequence outside caves in river valleys, those consist of a fining-up sequence [38, 39], similar vas found in the Sophie' s Cave: the basal gravel layer is followed by sands, by silts and and finally by clay, which the latter can often dry exposing mud cracks on the surface as found in the Sand Chamber (cf. Fig. **7**).

The sequences are partly incomplete in several areas of the Sophie's Cave, resulting in erosion within the next younger floods (cf. Chapter 4), but are generally present and excavated with a preserved trench in the Sand Chamber (Figs. **6, 7**), with coarse sand-clay short sequence sets [13]. 1-20 cm thin clay beds also expose drying sediment structures in cross-section, such as mud cracks and mud crack bowl structures (Figs. **7A, B**) [13], being typical in flood sediment clays, which can be compared to clay deposits along river banks outside caves [38, 39]. Such mud cracks in clay have also been figured in the "Middle Pleistocene" red clay deposits of the Zoolithen Cave, which were casted (hyporeliefs) below the first larger speleothem generation [14]. In one sand/clay layer of the yellow series of the Sophie's Cave, even flow structures are present (Fig. **7C**).

The complete yellow series sequence reflects well the climatic changes [38, 39],

10 cm

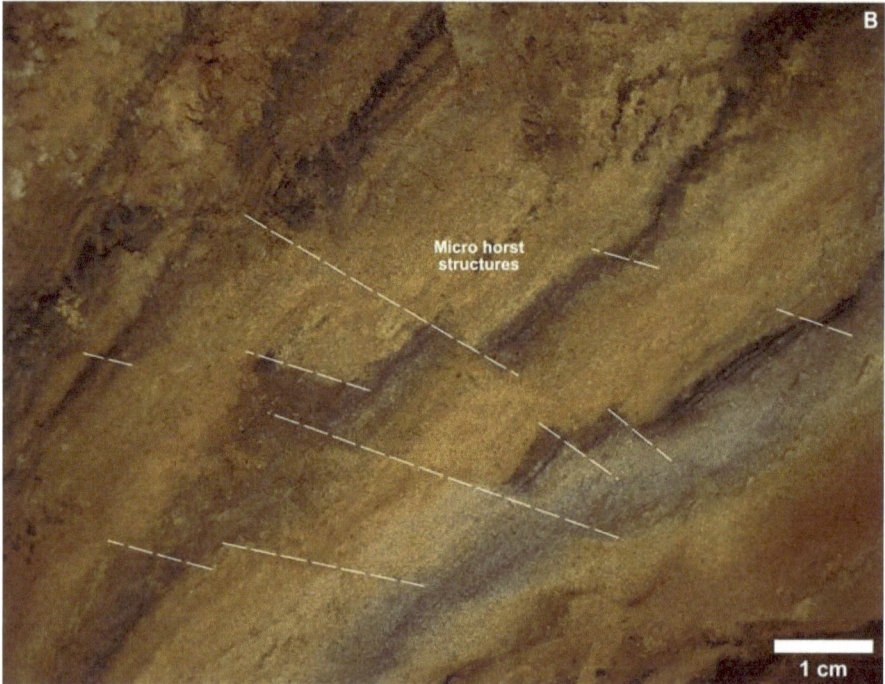

1 cm

Figure 4. A. Sand Chamber, Sophie's Cave, lower part of the section (section H), also refilled with the "colored series" of Pliocene/Early Pleistocene and finally also Middle Pleistocene ("yellow series"). **B.** Detail of the step-like micro horst structures, which do not reach into the overlaying Middle Pleistocene yellow sand series (see below, and Fig. **6**).

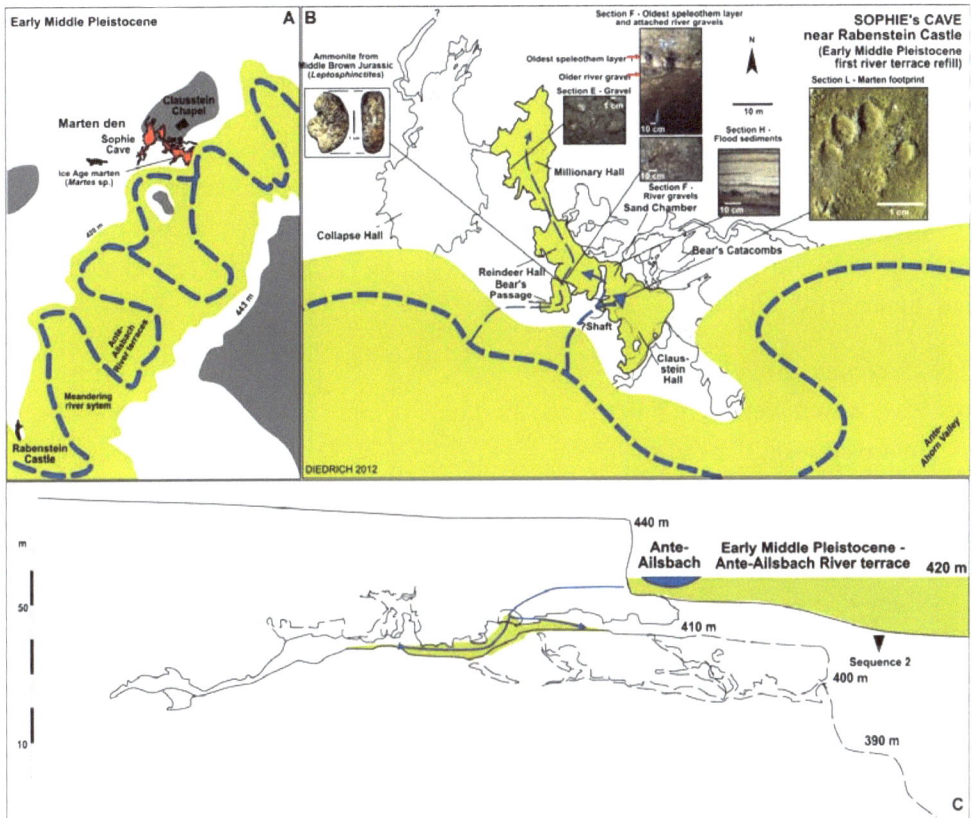

Figure 5. Middle Pleistocene (app. 780.000-200.000 BP) - Ante-Ailsbach River terrace (app. 420 m a.s.l.). **A.** Valley reconstruction **B.** Sophie's Cave map in dorsal view with refilled areas and finds including the new *Martichnus desseri* nov. ichnog. and ichnosp. marten track. **C.** Cave, terrace and filled areas in cross section (section outlines after T. Striebel) (notice the valleyward 3° angeled cave system tectonics) (modified after [13]).

especially the ground water elevations and humidity within the Middle Pleistocene in this region, and finally the change from a warm-period meander to a cold period braided river system in general [38, 39]. The Ante-Ailsbach River seems to have transported first by a meandering warm period stream, then during high water stands, sediments flowed into the cave from above the Reindeer/Clausstein halls. This cave was partly humid, partly dry, with the ground water already in the deeper cave levels (Figs. **1** and **5**). Also, this indicates, the river terrace was at that time at 425-420 m a.s.l., therefore, not even the first larger entrance of the Bear's Passage (see Chapter 4) was opened or accessible for cave animals. After this second refill, which was the first river terrace sediment input, the Ante-Ailsbach River terrace dropped in some meters elevation [13].

On the upper part of this "first terrace sequence" (yellow series) of the Ante-Ailsbach River terrace (completely eroded within the valley and present only with relic sediments in the Sophie's Cave, and possibly others along the valley), homogenous and up to several meters thick coarse river gravels (gavel sizes 1-5 cm) appear (Fig. **6**) [13].

These coarse gravels consist of one cm small, clean quartze and sandstone pebbles, which were originally deposited on the top of the Franconia Mesozoic Mountains during the Lower Cretaceous [1, 15, 40]. They were finally reworked obviously again in the Plio-Pleistocene. Most components are the "Hornstein concretions" which eroded from the White Jurassic reef dolomite rocks [1, 40].

Other abundant gravel components are yellowish silt/sandstone rocks, which originate from marine sediments of the Middle or "Brown" Jurassic [40, 41] (see Chapter 2, Fig. **1**), but must have been transported by the river from the Kirchahorn direction downstream [13]. At this time, the river quickly eroded deeply into the Middle Jurassic Ornaten Claystone (or "Oxford Clay") and Iron Oolithic sandstones (see Chapter 2, Fig. **1**) [13]. Clear original Jurassic age indication comes from those gravels with an ammonite steinkern fragment of *Leptosphinctes* sp. (Chapter 2, Fig. **1**), which originated from the Bathonian, Middle Jurassic [41]. The find also indicates that around Kirchahorn the saddle structure was only half eroded as today, and did not reach into the very soft dark Opalinus Claystones of the Early (or "Black") Jurasic layers (Chapter 1, Fig. **1**) [13]. Furthermore, the Middle Pleistocene gravels are built up by fossils originating from the Late Jurassic sponge reef dolomites. Fossil finds from those rocks are mainly silificated sea urchin spines and silificated spongae, the latter of which are mostly in hornstein concretions (cf. Chapter 2, Fig. **3**).

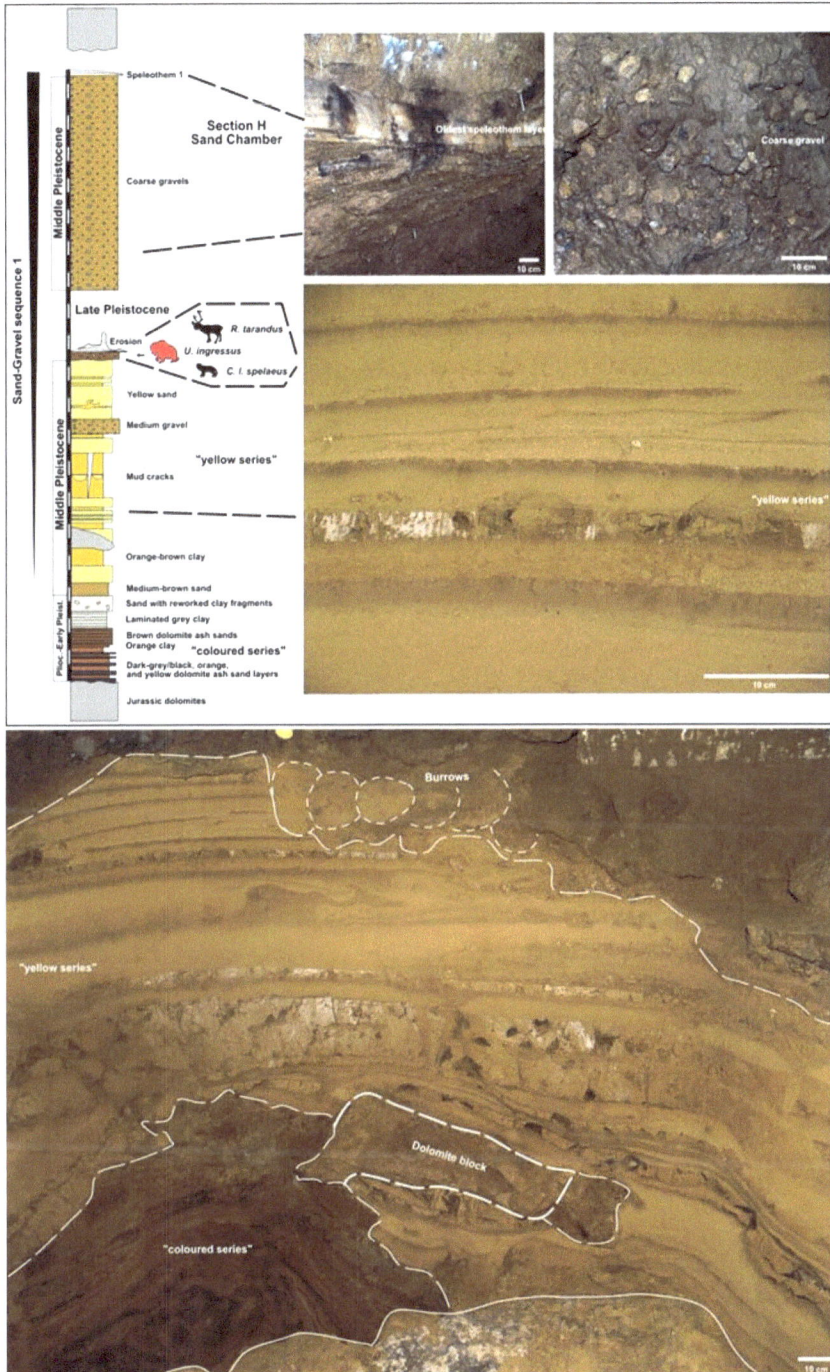

Figure 6. A. Section of the Sand Chamber, Sophie's Cave, with Middle Pleistocene river terrace high flood sediments from the Ante-Ailsbach Stream. **B.** Oldest speleothem layer covering. **C.** Coarse gravels (both photos upper Reindeer Hall). **D.** The yellow clay, sand and gravel sequence sediments in the Sand Chamber, the so-called "yellow series" of the first terrace sequence lies discordant and erosive on older, "coloured series"(modified after [13]).

The Pleistocene fluvial reworked gravels, originating of Middle Jurassic marine sediments were covered by the oldest non-absolute dated, about 10-30 cm thick speleothem layer (speleogenesis phase 1, Fig. **6**), which demonstrates a "warm period" (?Holstein Interglacial).

Possibly it correlates in age similar as known for main speleogenesis periods of other German/European caves, such as in the Harz Mountain Baumann's Cave and Unicorn Cave, or in the Upper Franconia Zoolithen Cave [14, 42 - 44].

This speleothem layer is attached as a relic after speleothem floor collapses due to fluvial sediment erosion below to many cave walls in different chambers, passages and halls, and is the best marker to measure the exact elevation of the original refill height of the Middle Pleistocene yellow series (Fig. **6**) [13].

In many places, in the Sand Chamber, Reindeer Hall or Millionary Hall and Bear's Passage, those gravels are represented mostly only in relics being attached below the oldest speleothem layer. Following, the speleothem layer and gravel elevation and distribution of the Reindeer Hall and Millionary Hall, those must have originally filled half way in elevation with sediments [13].

Their relics are even above the level of the top of the large Millionary stalagmite, which could not have been present therefore before the Middle Pleistocene. The most distal yellow series sediments are only sands and silts or clays which reached the passage to the Collapse Hall and even few into the North Passage (see section correlations in Fig. **1**) [13].

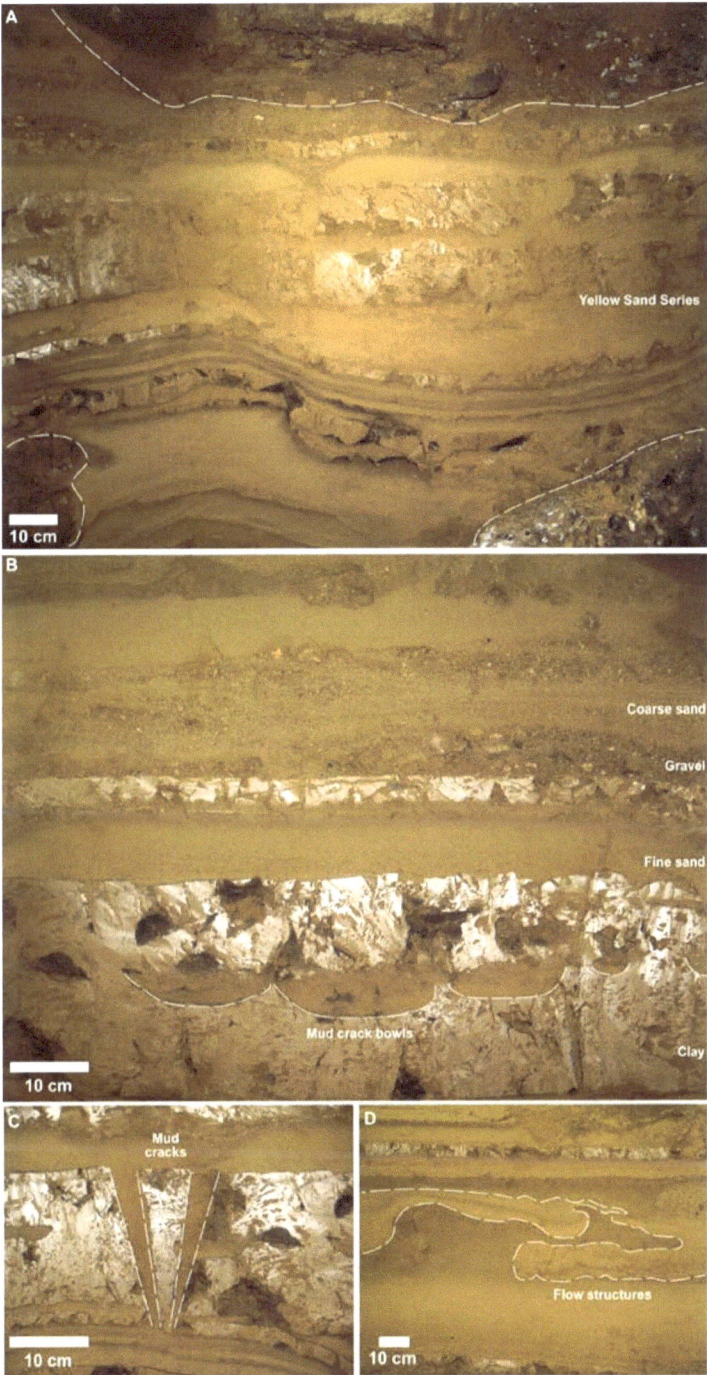

Figure 7. Details of the "yellow series" sediments and sediment structures in the Sand Chamber, Sophie's Cave, of the Middle Pleistocene river floodings of the Ante-Ailsbach Stream. **A.** The yellow clay, sand, and gravel layers contain: **B.** Mud cracks and **C.** Flow structures (extended after [13]).

The Oldest Middle Pleistocene Marten Footprints – First Small Carnivore Den

Quite unique are the German/European oldest "Middle Pleistocene" footprints of a small carnivore (Figs. **5B, 7**). The tracks were discovered in the Clausstein Hall. They are preserved as "casts" below the oldest speleothem layer, therefore being dated to be of Middle Pleistocene age, possibly before the (?Holsteinian) Interglacial (= first speleothem genesis period). The tracks, three incomplete of one trackway and two single footprints (Figs. **8B-D**) are preserved as hyporelief, were left originally printed on the mud crackedred clay surface that must have dried (and therefore preserved the tracks) and was finally covered by the oldest speleothem layer (Fig. **8A**). Those imprints are described and taxonomically named herein.

By its small footprint size of about 2.5 cm in length (heal pad to digit III), which have five digit pad impressions in front of a bean-shaped heal pad impression (Figs. **8B-C**), those fit well by comparison only to modern marten (*Martes*) tracks [45]. After comparing the recently reviewed Pleistocene European track record and ichnotaxonomy, which was missing marten tracks [46], the Pleistocene tracks must receive a new name.

For fossil marten tracks herein the new ichnogenus *Martichnus* nov. ichnogen. is introduced, whereas the only known ichnospecies yet from the Middle Pleistocene (?Holsteinian) is called *desseri* nov. ichnosp. (in memorial after Mr. W. Dess). The ichnoholotype *M. desseri* (Fig. **8A-C**), which is not clear to determine as fore or hind left foot imprint, is *in situ* below the speleothem slab in the Clausstein Hall (= within public protected natural monument).

Important are the marten track finds for the Sophie's Cave genesis and geomorphology reconstruction, because it is the oldest indirect small carnivore den on record of Upper Franconia – a possible "stone marten" cave den. The

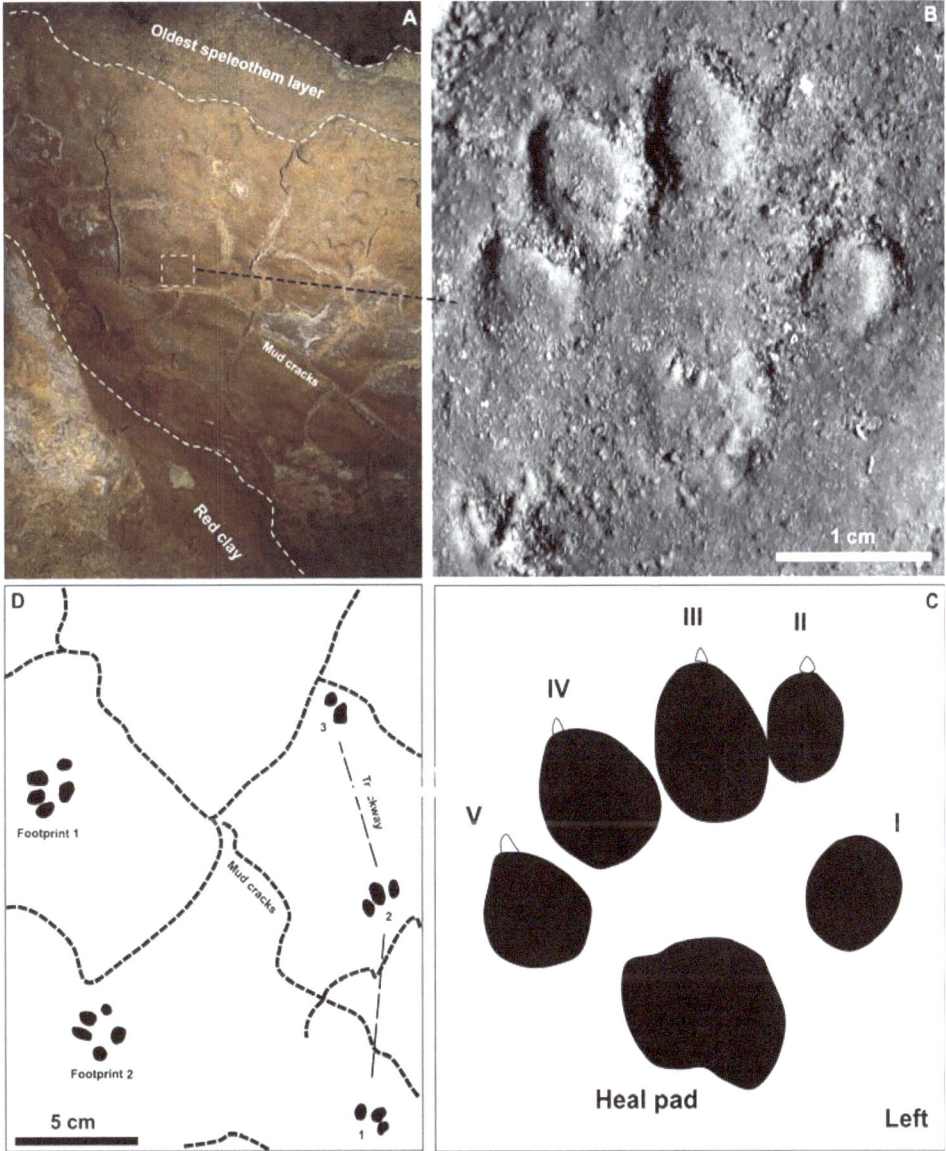

Figure 8. Marten tracks of the ichnotype *Martichnus desseri* nov. ichnogen. nov. ichnospec. from the Clauststein Hall, Sophie's Cave. **A.** Those are preserved on the bottom (hyporelief) of the oldest speleothem generation, which has also casted the mud cracks of the underlaying red clay. **B-C.** Most complete and ichnoholotype footprint with five digit pad and heal impression. **D.** At minimum three incomplete trackways are mapped, two consisting of each a single footprint, and one of three footprints of which only the middle digits are printed deeply.

marten remains from the Zoolithen Cave and some other caves of Upper Franconia [47, 48] are not well described yet, and comprise cranial records at minimum [47, 48], which age also might reach into the Middle Pleistocene [14]. Most probably it was the stone marten *Martes foina* (Erxleben 1777), which also use modern caves as dens [49]. This also indicates that before the Interstadial (?Holsteinian), the cave became dry in its upper parts (mud cracked red clay bed below oldest speleothem layer) after the first massive river terrace fills (yellow series) and was used first by small carnivores, which could have penetrated only via the small vertical shaft into the Clausstein Hall, before it was closed finally completely by the first speleothem generation.

LATE MIDDLE PLEISTOCENE (?200.000-113.000 BP) – MASSIVE EROSION

With further lowering of the ground water due to climate change and further tectonic uplift of Upper Franconia due to the Bohemian Massif uplift since Pliocene times [34], the Ante-Ailsbach River terrace became narrower and deeper (elevation about 415 m a.s.l.). At this period, massive new valley-sided river terrace floods removed several meters of the Middle Pleistocene refill (yellow series), and caused a massive collapse and erosion of the first speleothem generation, which also gravitatived deeper into unknown cave levels [13]. That massive sediment erosion opened and connected the halls, passages and chambers again (Fig. 1). The new flood events can be dated by the stratigraphy and with the oldest cave bear subspecies from the Bear's Passage, Reinder and Millionary halls (cf. Chapter 4), which seem to not be older than 113.000 BP, indirectly to have happened before the Late Pleistocene.

REFERENCES

[1]   Gümbel CW von. Geognostische Beschreibung des Königreichs Bayern. Band 4: Geognostische Beschreibung der Fränkischen Alb (Frankenjura) mit dem anstossenden Fränkischen Keupergebiete. Cassel, Fischer 1891; p. 761.

[2]   Heller J. Muggendorf und seine Umgebungen oder die fränkische Schweiz. Ein Handbuch für Wanderer in diese Gegend, mit den Reiserouten und nothwendigen Notizen für Reisende. Neue, sehr vermehrte Auflage. Mit einer Charte. Bamberg: Dresch 1842; p. 234.

[3]   Neischl A. Die Höhlen der fränkischen Schweiz und ihre Bedeutung für die Entstehung der dortigen Täler. Nürnberg, Schrag 1904; p. 95.

[4]   Brückner K. Führer durch die fränkische und Hersbrücker Schweiz. 2 Auflage. Wunsiedel, Kohler, 1907; p. 296.

[5]   Habbe K-A. Der Karst der Fränkischen Alb - Formen, Prozesse, Datierungsprobleme. Die Fränkische Alb. Schr Zentr Inst fränk Landesk Univ Erlangen 1989; 28: 35-69.

[6]   Meyer RKF, Schmidt-Kaler H. Wanderungen in die Erdgeschichte (5). Durch die fränkische Schweiz. München: Pfeil-Verlag 1992; p. 167.

[7]   Kaulich B, Schaaf H. Kleiner Führer zu den Höhlen um Muggendorf. Nürnberg 1993; p. 125.

[8]   Baier A. Hydrogeologie Frankens: Heilwässer, Wasserstollen und Karstquellen. Jb mittelrh geol Ver N F 2003; 85: 95-167

[9]   Heller J. Muggendorf und seine Umgebungen oder die fränkische Schweiz. Ein Handbuch für Wanderer in diese Gegend, mit den Reiserouten und nothwendigen Notizen für Reisende. Bamberg: Dresch 1829; p. 215.

[10] Cave cadastre FHKF (unpublished documents).

[11] Sieghardt A. Im Bannkreis der Wiesent. Kultur, Geschichts- und Landschaftsbilder aus der Fränkischen Schweiz. Band 1 Nürnberg. Koch 1925; p. 203.

[12] Buckland W. Reliquiae Diluvianae, or observations on the organic remains contained in caves, fissures, and diluvial gravel, and other geological phenomena, attesting the action of an universal deluge. London: J. Murray 1823; p. 303.

[13] Diedrich C. Ice Age geomorphological Ahorn Valley and Ailsbach River terrace evolution– and its importance for the cave use possibilities by cave bears, top predators (hyenas, wolves and lions) and humans (Late Magdalénians) in the Frankonia Karst – case studies in the Sophie's Cave near Kirchahorn, Bavaria. Quat Sci J 2013; 62(2): 162-74.

[14] Diedrich C. Holotype skulls, stratigraphy, bone taphonomy and excavation history in the Zoolithen Cave and new theory about Esper's "great deluge". Quat Sci J 2014; 63(1): 78-98.

[15] Groiss JT, Kamphausen D, Michel U. Höhlen der nördlichen Fränkischen Alb: Entwicklung, Fauna, Karst-Hydrologie. Erlanger geol Abh Sonderbd 1998; 2: 161-8.

[16] Bretz JH. Vadose and phreatic features of limestone caverns. J Geol 1942; 50: 675-811.

[17] Ford DC, Williams PW. Karst geomorphology and hydrology. London: Unwin-Hyman 1989; p. 601.

[18] Klimchouk AB. Hypogene Speleogenesis: Hydrogeological and Morphogenetic Perspective. National Cave and Karst Research Institute. Carlsbad NM Spec Pap 2007; 1: 1-106.

[19] Jennings JN. Karst Geomorphology. Oxford: Blackwell 1985; p. 293.

[20] Dogwiler T, Wicks CM. Sediment entrainment and transport in fluviokarst systems. J Hydrol 2004; 295: 163-72.

[21] White WB. Cave sediments and paleoclimate. J Cave Karst Stud 2007; 69(1): 76-93.

[22] Sasowsky ID, Mylroie J. Studies of Cave Sediments Physical and Chemical Records of Paleoclimate. 2nd ed. Stuttgart-Heidelberg-New York, Springer-Verlag GmbH 2007; p. 329.

[23] Schabdach H. Die Sophienhöhle im Ailsbachtal. Wunderwelt unter Tage. Ebermannstadt: Verlag Reinhold Lippert 1998; p. 47.

[24] Freyberg von B. Tektonische Karte der Fränkischen Alb und ihrer Umgebung. Erlanger geol Abh 1969; 77: 1-81.

[25] Brand F. Was können wir aus lehmigen Ablagerungen der Binghöhle ablesen? In: Brand F, Illmann R, Leja F, Preu D, Schabdach H. (Eds.), Die Binghöhle bei Streitberg – Auf den Spuren eines unterirdischen Flusses. Streitberg 2006; pp. 28-34.

[26] Heller F. Eine Forest-Bed-Fauna aus der Sackdillinger Höhle (Oberpfalz). N Jb Min Geol Paläont B Beil-Bd 1930; 63: 247-98.

[27] Spöker RG. Die jungpliozänen Ablagerungen in der Sackdillinger Höhle und ihre Beziehungen zur Landschaft. Ein fossiler Wasserschlinger. N Jb Min Geol Paläont B Beil-Bd 1933; 70: 215-26.

[28] Brunner G. Eine präglaziale Fauna aus dem Windloch bei Sackdilling (Oberpfalz). N Jb Min Geol Paläont B Beil-Bd 1933; 71: 303-28.

[29] Brunner G. Das Fuchsloch bei Siegmannsbrunn (Oberfr.), eine mediterrane Riss-Würm-Fauna. N Jb Min Geol Paläont Abh 1954; 100: 83-118.

[30] Gibbard PL, Cohen KM. Global Chronostratigraphical correlation table for the last 2.7 million years. Episodes 2009; 31(2): 243-7.

[31] Goldfuss GA. Die Umgebungen von Muggendorf. Ein Taschenbuch für Freunde der Natur und Alterthumskunde. Erlangen 1810; p. 351.

[32] Burger D. Dolomite weathering and micromorphology of paleosoils in the Franconian Jura. Catena Supplement 1989; 15: 261-7.

[33] Davis KJ, Dove PM, De Yoreo JJ. The role of Mg2+ as an impurity in calcite growth. Science 2000; 290: 1134-7.

[34] Bankewitz P, Bankewitz E, Bräuer K, Kämpf H, Störr M. Deformation structures in Plio- and Pleistocene sediments (NW Bohemia, Central Europe). Geol Soc London Spec Publ 2003; 216: 73-93.

[35] Spöker RG. Zur Landschafts-Entwicklung im Karst des oberen und mittleren Pegnitz-Gebietes. Remagen, Verlag des Amtes für Landeskunde 1952; p. 60.

[36] Diedrich C. Impact of the German Harz Mountain Weichselian ice-shield and valley glacier development onto Palaeolithics and megafauna disappearance. Quat Sci Rev 2013; 82: 167-98.

[37] www.br.de/nachrichten/oberfranken/inhalt/erdbeben-tschechien-oberfranken-100.html

[38] Kaiser K. Gliederung und Formenschatz des Pliozäns und Quartärs am Mittel- und Niederrhein sowie in den angrenzenden Niederlanden unter besonderer Berücksichtigung der Rheinterrassen. Festschrift Deutscher Geographen-Tag, Köln. Wiesbaden 1961; pp. 236-78.

[39] Bridgland DR, Maddy D, Bates M. River terrace sequences: templates for quaternary geochronology and marine-terrestrial correlation. J Quat Sci 2004; 19(2): 203-18.

[40] Freyberg von B. Zur Stratigraphie und Fazieskunde des Doggersandsteins und seiner Flöze. Geol Bavarica 1951; 9: 1-108.

[41] Schlegelmilch R. Die Ammoniten des süddeutschen Doggers. Stuttgart, Gutsav-Fischer-Verlag 1985; p. 284.

[42] Dorale JA, Edwards RL, Alexander Jr EC, Shen C-C, Richards DA, Cheng H. Uranium dating of speleothems: Current techniques, limits and applications. In: Sasowsky ID, Mylroie JE. (Eds.) Studies of cave sediments. New York: Kluwer-Academic/Plenum 2004; pp. 177-226.

[43] Richter DK, Götte T, Niggemann S, Wurth G. REE3+ and Mn2+ activated cathodoluminescence in late glacial and Holocene stalagmites of central Europe: evidence for climatic processes? The Holocene 2004; 14: 759-67.

[44] Kempe S, Rosendahl W, Wiegand B, Eisenhauer A. New speleothem dates from caves in Germany and their importance for the Middle and Upper Pleistocene climate reconstruction. Acta Geol Polon 2002; 52(1): 55-61.

[45] Bang P, Dahlström P. Tierspuren. BLV-Naturführer München 2000; p. 263.

[46] Diedrich C. Tracking Late Pleistocene steppe lion predation on its guild in a mammoth steppe near a river bank spotted hyena open air den of Bottrop, Northwestern Germany. (in review).

[47] Groiss JT. Paläontologische Untersuchungen in der Zoolithenhöhle bei Burggeilenreuth. Ein vorläufiger Bericht. Erlanger Forsch B Naturwiss 1979; 5: 79-93.

[48] Eberlein C. Die Musteliden aus drei Höhlen des Frankenjura (Zoolithenhöhle, Geudensteinhöhle und Höhle bei Hartenreuth). Unpublished diploma-thesis. University Erlangen 1996.

[49] Herra J, Schley L, Engel E, Roper TJ. Den preferences and denning behaviour in urban stone martens (*Martes foina*). Mam Biol, Z Säugetierk 2010; 75(2): 138-45.

CHAPTER 4

THE EARLY/MIDDLE LATE PLEISTOCENE CAVE BEAR DEN

Abstract: With the beginning of the early Late Pleistocene glacial period (or even earlier: ?late Saalian/Eemian) and Ailsbach terrace elevation 415 m a.s.l., small cave bears penetrated only a side branch (Bear's Passage) the Sophie's Cave and used it as den. Nine cave bear nests *Ursalveolus carpathicus* Diedrich 2011 were documented with larger to medium-sized round-oval depressions in the deepest cave bear den part of the Millionary Hall. Autochthonous cave bear skeletal parts, especially partly connected vertebral columns were found in all den areas of the Bear's Passage, Reindeer Hall bone field and the Millionary Hall, partly being in place. A systematic excavation of the bone field, which was left in the cave *in situ* (also for visitors) demonstrate a mainly adult population within this hall. Using a combination of the skull shape morphology, P4 tooth morphology and C^{14} dated teeth from other German/European cave bear dens, the small cave bears of those cave areas can be identified as small cave bears *U. spelaeus eremus/spelaeus* Rabeder *et al.* 2004. A new composite skeleton including male/female adult-senile bone material from different individuals was arranged for "*U. s. cf. eremus*" which is presented in a show case within the cave, which is anatomically nearly complete including in Europe unique for a "skeleton" all nine "tongue bones". These small cave bears also being known from the nearby Große Teufels Cave, Zoolithen Cave inhabited in Upper Franconia the Sophie's Cave and other caves between approx. 113.000-32.000 BP dated biostratigraphically with the P4 tooth morphology. With the P4 morphotypes blocking events can be coarsely estimated, whereas most primitive three-coned forms appear in the deeper Millionary/Reindeer halls and Bear's Passage. Only in the latter higher evolved forms demonstrate a longer use of this branch which was blocked to the Reindeer Hall most probably during an interstadial (possibly around 42.000 BP). At the end of the middle Late Pleistocene, finally the former still unknown entrance was also blocked, which did not allow smaller cave bears to use the cave as a den anymore.

Keywords: Early/Middle Late Pleistocene, sedimentology, terrace gravel infill, Ailsbach Valley geomorphology, small cave bear species/subspecies, cave bear clock, cave bear den, hibernation nests, bone taphonomy, scavengers, cave bear pathology.

Cajus G. Diedrich

EARLY/MIDDLE LATE PLEISTOCENE (113.000-32.000 BP, MIS 5D-3) – FIRST CAVE BEAR DEN

At the beginning of the Late Pleistocene the Ailsbach River terrace was still at a high elevation, but the first small cave bears were able to enter the cave for denning [1]. At that time, the today's entrance was still closed/non-eroded and in elevation several meters below the terrace.

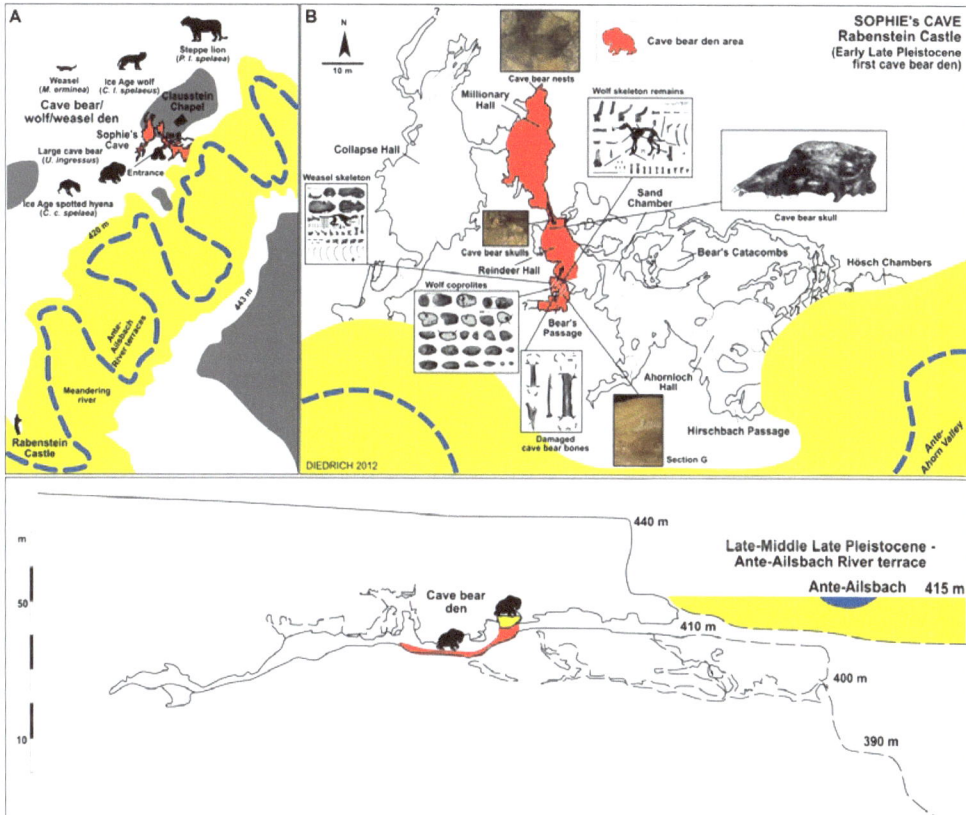

Figure 1. Early/Middle Late Pleistocene (113.000-32.000 BP, MIS 5d-3) – first small cave bears *Ursus spelaeus* cf. *eremus* Rabeder *et al.* 2004. The Ante-Ailsbach terrace was about 415-417 m a.s.l.. Cave bears entered the cave in what is today's blocked, former medium-sized entrance. The Reindeer and Millionary halls were used for cave bear hibernation and for the birth of their cubs (modified from [1]).

The first accessible and larger entrance for cave bears was at the "beginning" of the Bear's Passage, which entrance remains still blocked [1]. This passage was filled with unknown in thickness (at minimum 1.5 meters thick) bone-rich sediments (= bonebed in dolomite ash sand) [1]. The early smaller cave bears were able to enter the Sophie's Cave at an elevation around 415 m a.s.l. into the Bear's Passage only, and from there, deeper through the 1,5 meter wide opening at the end of the passage deeper into the Reindeer Hall [1]. The deepest accessible area was the Millionary Hall (Passage to the Collapse Hall, Fig. **1**) [1].

The Small Cave Bear Subspecies Composite Skeleton "Benno"

In 2011, the historically composed "cave bear" skeleton mounted in the Reindeer Hall for visitors, which was under extremely bad condition covered by algae, was demounted and analyzed to separate its original and casted bones.

The bones (*e.g.* skull, several ribs, vertebrae, limb bones) originated from the knowledge gained of the new cave and first cave bear bone and tooth studies [1] from a European DNA-tested and well-known small cave bear subspecies of *Ursus spelaeus* cf. *eremus* Rabeder *et al.* 2004 ([2 - 9], Figs. **1 - 6**). The "Benno skelton" also incorporated the large cave bear species *U. ingressus* Rabeder *et al.* 2004 [2 - 9] (see Chapter 6) from the anterior cave areas (Ahornloch, Claustein Halls), which were mixed up with casted sternal bones and intercostal cartilage replica. The skull included two composed right lower jaw mandibles of the larger cave bear species. The newly composed skeleton (by PaleoLogic), which is based anatomically on a modern brown bear skeleton [10], has less than 25% of the formerly used "Benno skeleton" bones included and contains now only bones of grown up individuals which were found in the Bear's Passage (selected from about 1.342 bones of the "bone dump"), and a few being from the lower Reindeer Hall area (selected from few large bones of the former Rabenstein Castle collection, rediscovered in 1998).

The small cave bear *Ursus spelaeus* cf. *eremus* Rabeder *et al.* 2004 (Figs. **2 - 6**) comp-osite skeleton presented herein includes male/female bones, which postcranial bones are not well analysed yet from European sites in their proportion osteometric differences, especially postcranial bones. This composition is unique at the moment in Europe and includes, in contrast to most "cave bear" skeleton composites from Europe, the small sternal bones, sesamoid bones and "tongue

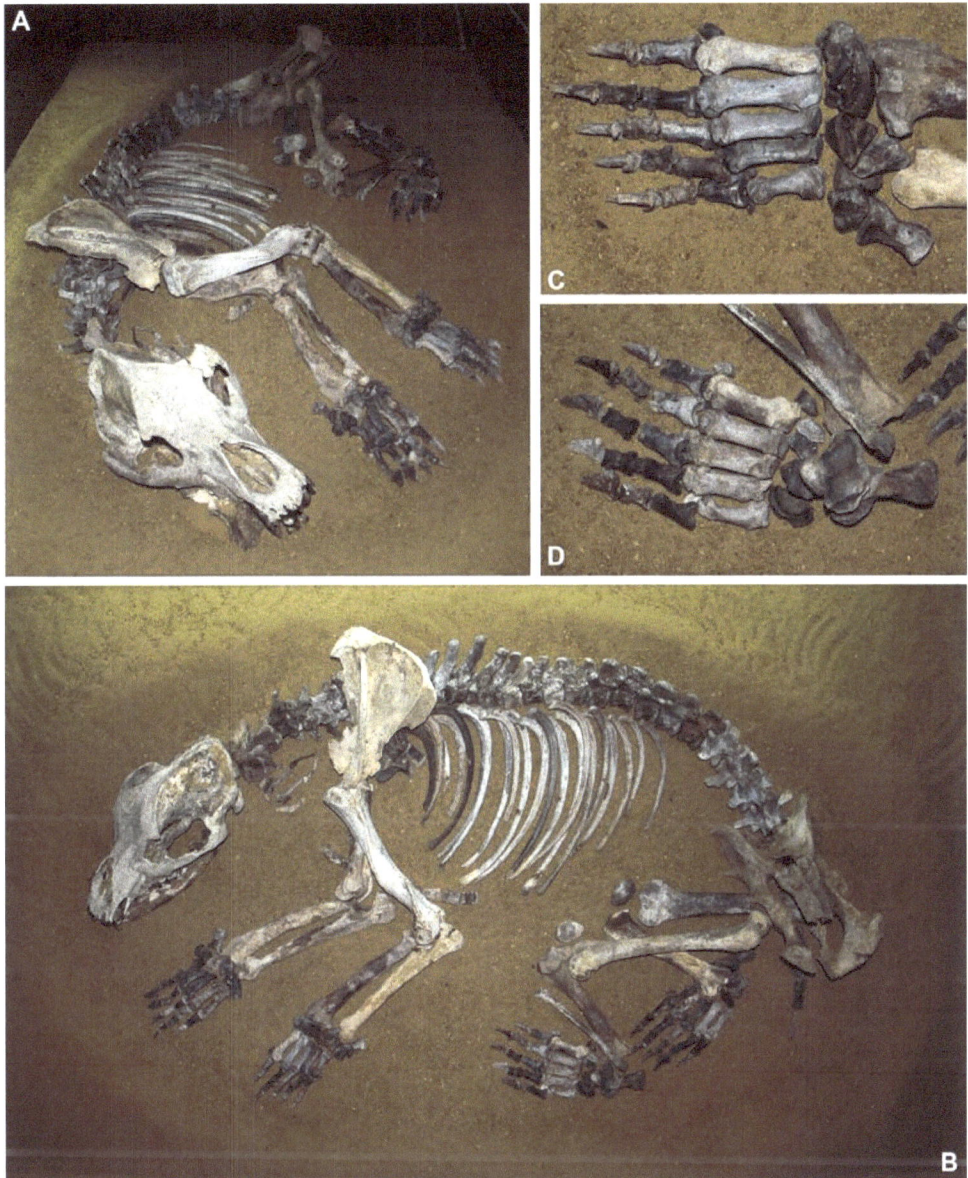

Figure 2. The Early/Middle Late Pleistocene small cave bear *Ursus spelaeus* cf. *eremus* Rabeder *et al.* 2004 "Benno" composite skeleton after the new composition by PaleoLogic in 2011. All bones are mainly from the Bear's Passage, with a few ones from the Reindeer Hall, Sophie's Cave (coll. Rabenstein Castle Museum, in a new show case within the cave). **A.** cranial view. **B.** Skeleton in hibernation position in a cave bear nest in lateral view. **C.** Left manus skeleton. **D.** Left pes skeleton.

Figure 3. The small cave bear *Ursus spelaeus eremus* Rabeder *et al*. 2004, with flattened skull and back sloping profile Reconstruction drawing of the cave bear composite skeleton "Benno" (Illustration G. "Rinaldino" Teichmann).

bones". Most of the bones used for the comp-osite are complete, except the right humerus and middle ribs, which are missing distal parts, and the thorax, which lacks some exact fitting costae.

The skull of an older adult animal with stronger tooth use (Fig. **4**) is reused from the former skeleton composite, but three teeth including the right canine, which were obtained by sediment sieving from the Bear's Passage fit 100% in the unique, finger-print-like formed alveolar groove shapes (Fig. **4**). The skull can be therefore securely relocated to the original place where it was found. For the Eurropean record it is unique to have the integration of all nine "tongue bones", one median basihyoid, and each right and left two ceratohyoids, thyrohyoids, epihyoids, and stylohyoids (Fig. **4**), which were selected from the bone dump material of the Bear's Passage.

Such smaller bones are always missing in mounted cave bear skeletons of *e.g.* the Upper Franconia Zoolithen Cave or the Große Teufels Cave or other caves in Germany [1, 11, 12].

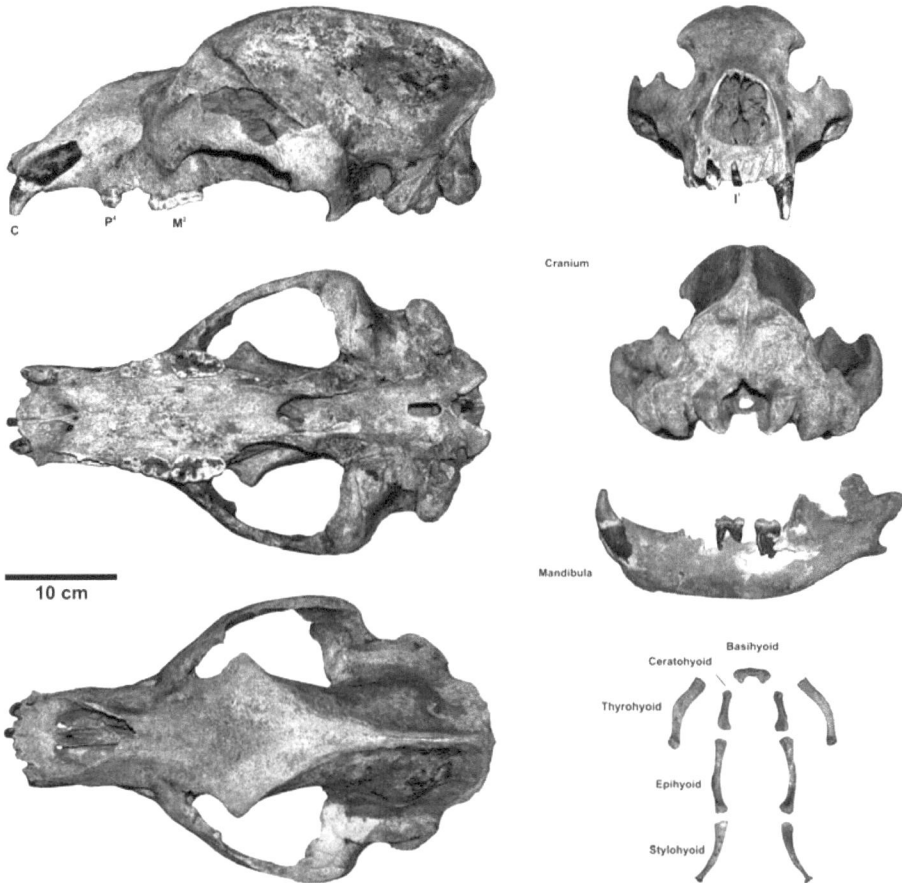

Figure 4. The small cave bear *Ursus spelaeus* cf. *eremus* Rabder *et al.* 2004. Skull in different views, and tongue bone apparatus consisting of nine bones (one basihyoid, and each two ceratohyoids, thyrohyoids, epihyoids and stylohyoids), all found at the end of the Bear's Passage, Sophie's Cave (coll. Rabenstein Castle Museum, "Benno" composite skeleton).

The postcranial axial skeleton is following the brown bear anatomy [10] and has 7 cervical, 14 thoracic, and 6 lumbar vertebrae. The exact amount of the caudal vertebrae remains unclear, but six are figured being in different sizes including the final one (Fig. **5**). The 14 ribs are in different shapes (Fig. **5**), and were able to select with correct anatomic positioned ones in most cases. Only the last two have single-head articulations. Seven sternal bones are composed (Fig. **5**), whereas only the first is secure in its position due to its cross-like shape.

The appendicular skeleton is figured only with the left side herein, with partly mirrored bones (Fig. **6**). The now replaced pelvis (Fig. **6**, selected from 1998 rediscovered Rabensein Castle collection material) of the former composite was from a large male of *U. ingressus* (being also figured in Chapter 6). The fore- and hind limb longbones (humerus, ulna, radius, femur, tibia and fibula) are also easy to select from the large cave bears, as also being demonstrated in strong proportion size differences of those different cave bears at other German cave bear dens such as the Hermann's Cave [12]. This is herein similar if comparing Fig. **6** and in Chapter 6, Fig. **5**, which are both in similar scales.

The forelimbs are composed of small and nearly identical in symmetrical sizes incomplete scapulae, but complete humeri, ulnae and radii (Fig. **6**). The manus skeleton is built after the larger pisiform and scapholunatum of the five smaller carpalia bones (Fig. **6**). The five digits are correctly in the metacarpus bones, even being nearly similar proportioned on the left and right side (Fig. **6**). Not secure are the phalanx I to III composites, also a result of the few available small pedal bone material from the bone dump. However, each has below the mc/phalanx I joints paired sesamoids (Fig. **6**). Whereas mc II-V have all a phalanx II, this is naturally absent in the digit I, where the phalanx III directly is attached to the phalanx I (Fig. **6**).

The hindlimb (Fig. **6**) is for both sides is also nearly complete with the longbones (femurae, tibiae, fibulae) and the patellae, but the right tibia only is incomplete, because no complete one was possible to obtain from the bone dump. Below the calcaneus and astragalus, five smaller tarsalia follow (Fig. **6**). Also here, all metatarsus bones are anatomically correct and on each side in most cases similar in length. Again, all phalanx I-III bones and sesamoids are composites and in their position not secure. As for the manus, in digit I only phalanx II is naturally absent (Fig. **6**).

Small Cave Bear Carcasses and Bones from the Bear's Passage

In the Bear's Passage, there is only one excavated and documented section preserved (not fully excavated or figured herein yet in detail) [1].

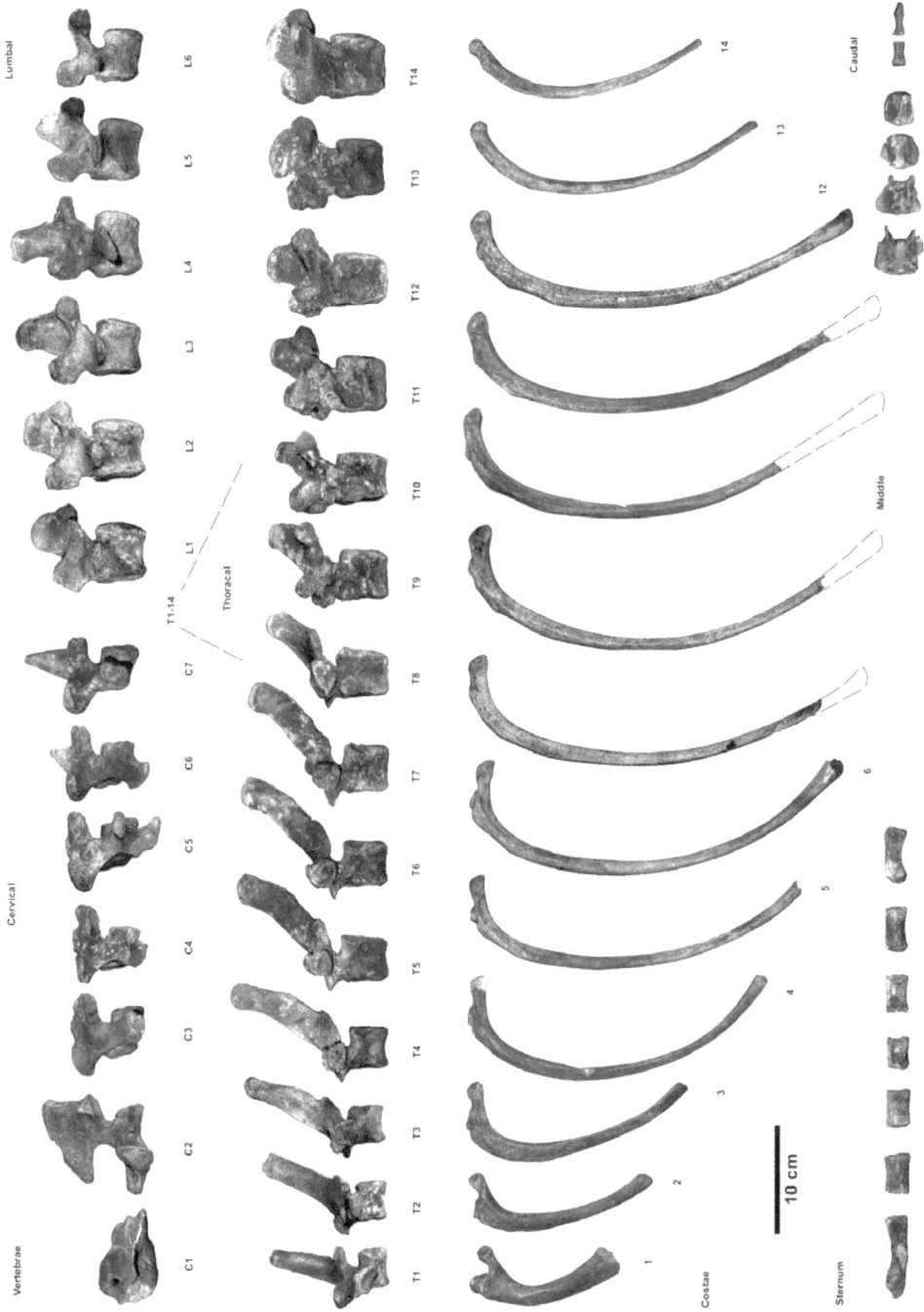

Figure 5. The small cave bear *Ursus spelaeus* cf. *eremus* Rabder *et al.* 2004. Axial skeleton (vertebrae, ribs, and sternal bones, all in lateral view, except caudal vertebrae in dorsal view, some ribs are mirrored) of the composite skeleton "Benno", with most of the material being from the Bear's Passage, and a few pieces possibly from the Reindeer Hall, Sophie's Cave (coll. Rabenstein Castle Museum, "Benno" composite skeleton).

This consists in its upper exposed part of redeposited (partly by former speleologists) yellow dolomite ash sands, a frost brekzia layer, and thin clay layers in which high amounts of bones and teeth of cave bears, some of wolves (including their coprolites), and finally a skeleton of a weasel were found (Fig. **1**) [3].

After ageneral cleaning of the disturbed sediments and washing of the historically compiled bone dump material (1.342 bones) at the end of the Bear's Passage, a small surface was excavated systematically close to the cave wall with mapping of all bones and coprolites (Fig. **7**), to understand the situation of the "old dumped" and skull, mandible or tooth slelected bone material" (resulting possibly in part to Neischl excavations [2], or younger illegal digs). As expected from the dump, cave bear bones and teeth were found from cubs and grown up animals, whereas many of the bones expose bite damages (see Chapter 5). Important are the *in situ* finds of wolve *Canis lupus spelaeus* Goldfuss 1823 bones or few teeth or their coprolites [3].

Another in place find is a *Mustela erminea* Linnaeus 1758 subsp. skeleton (see Chapter 5) that was found within its burrowed den (Figs. **7B, D**) [3]. Most of its bones were found after one millimeter in size of intensive sieving of the skeleton surrounding sediment. At least the skull and a lumbar vertebra were mapped in their in place situation (cf. Fig. **7B, D**).

From the "dolosands" of the Bear's Pasage several cave bear cub and milk teeth have been found within the "cleaning and sieving" (up to 1 mm) of the historically reworked sediments, which underline the dens used for cub raising purposes as well known for cave bear dens all over Europe [12 - 17]. However, there must be more sieving to obtain a better amount of the neonate cave bear remains in future, which seem to be mainly limited within the Bear's Passage, being absent in the deeper cave parts. Smallest bones of neonate cave bears result of death-births, or of few days survival, and often originate from initially articulated and finally scattered skeletons [12 - 18]. In raising cubs, it must be expected they will be spending the first few months in the cave, as is suggested from European cave dens [12 - 17].

Small Cave Bear Carcasses and Bones from the Reindeer Hall (Bone Field)

In total, 199 bones and teeth were excavated in systematic paleontological surface excavation technique style in the lower part of the Reindeer Hall (Figs. **8 - 9**). This area was covered by historic mud, and many holes were found within the heavily damaged sinter layer. Therefore it was decided, after the first discoveries below the sinter layer to remove also those fragments to obtain valuable scientific information and for visitor presentations interests "original *in situ*" bonebed situations. Each bone was left in-place after its discovery with correct orientation. Before mapping, and photographing such as study of the bite damages, those were cleaned in the cave with water. This caused a grey-black humid bone colour in the photos (Fig. **9**, light grey if dry).

This time-consuming process, the first systematic surface excavation for cave bears in a cave made in a Upper Franconia Cave since the too long "cave bear hunting" history [1, 13], was important for the bone and carcass taphonomy analyses (see Chapter 5). This bone field is another key site for the understanding of the cave use by cave bears not only for this locality. The analyses supported, that in this area, such as the Bear's Passage cave bear remains, are "autochthounous" and not "washed" into this cave area. This is different in the anterior Sophie's Cave with the large cave bear bonebeds (see Chapter 6). It is also different than the Zoolithen Cave [13] and unpublished information by own observations it is clear in the Große Teufels Cave, where the entire bonebeds are alloctonous - accumulated by water influence. Therfore, this bone field is a key for the understanding of the palaeoecology of cave bears. The detailed mapping of the bones of the bone field (Fig. **8**) demonstrates finds of cave bears in their hibernation areas and a medium disarticulation of their carcasses only by 1. Slow floating/sipping water impact, and 2. Top carnivores (see Chapter 5).

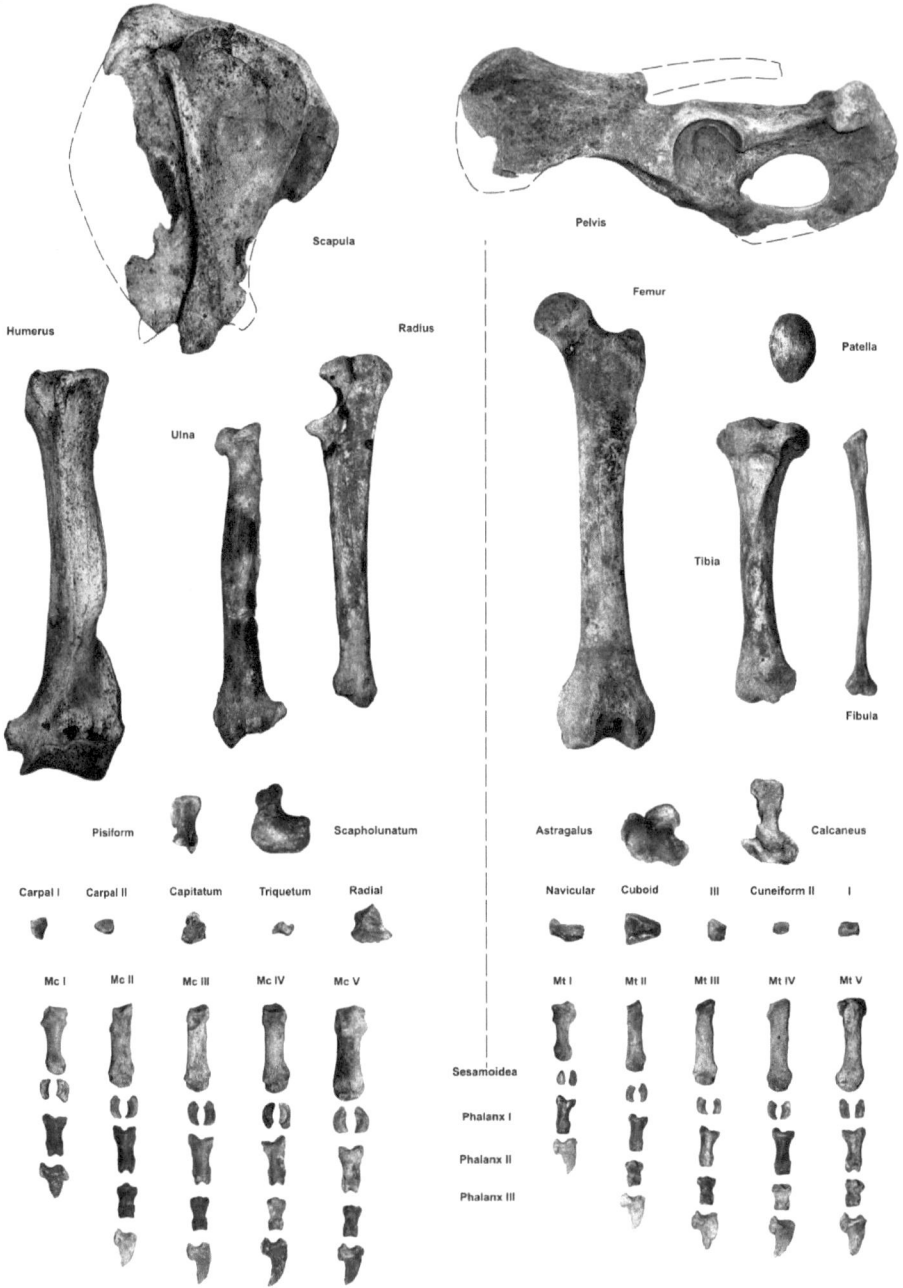

Left
fore limb

10 cm

Left
hind limb

Scapula

Pelvis

Humerus

Radius

Femur

Patella

Ulna

Tibia

Fibula

Pisiform

Scapholunatum

Astragalus

Calcaneus

Carpal I   Carpal II   Capitatum   Triquetum   Radial

Navicular   Cuboid   III   Cuneiform II   I

Mc I   Mc II   Mc III   Mc IV   Mc V

Mt I   Mt II   Mt III   Mt IV   Mt V

Sesamoidea

Phalanx I

Phalanx II

Phalanx III

Figure 6. The small cave bear *Ursus spelaeus* cf. *eremus* Rabeder *et al*. 2004. Left fore and hind limb bones in lateral or cranial views (pelvic mirrored), all found at the end of the Bear's Passage, Sophie's Cave (coll. Rabenstein Castle Museum, "Benno" composite skeleton).

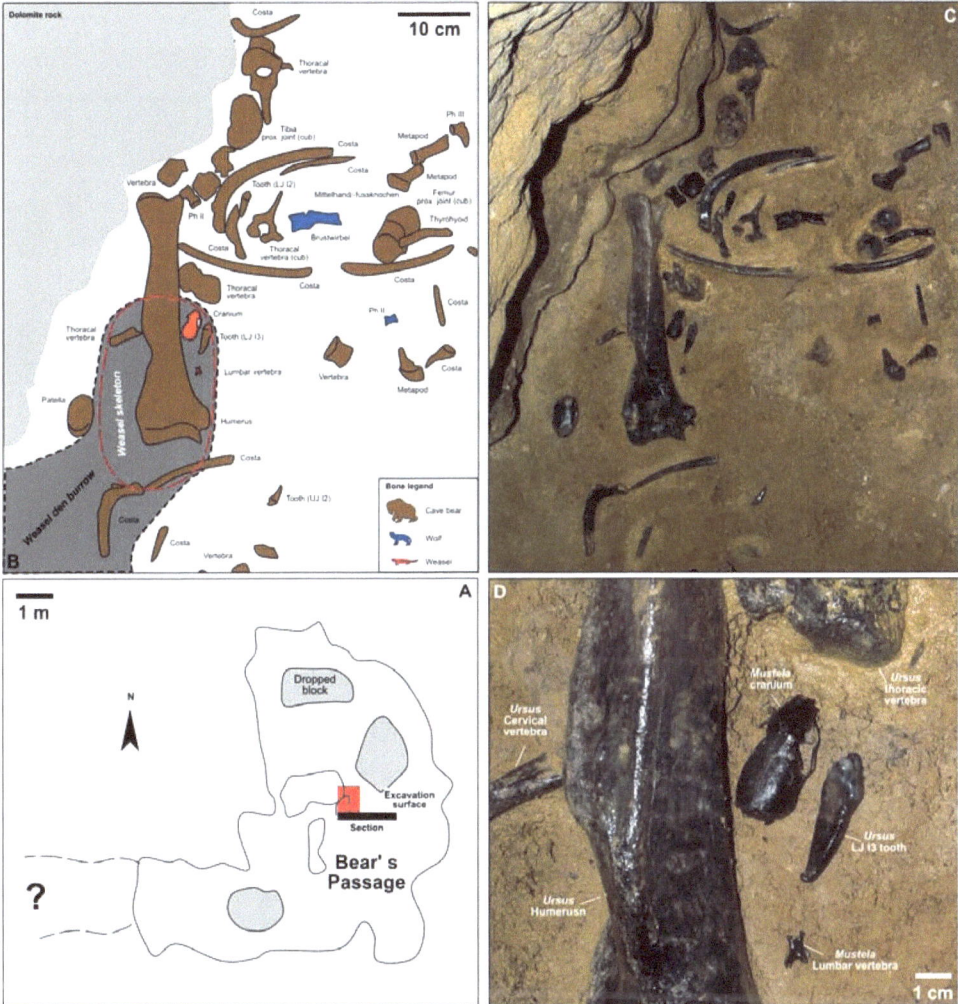

Figure 7. Bones of small cave bear *Ursus spelaeus* cf. *eremus* Rabeder *et al*. 2004, wolves, and weasels from the Bear's Passage, Sophie's Cave. **A.** Excavation surface. **B-C.** Systematically excavated surface. **D.** In an animal burrow (weasel den), a nearly complete weasel skeleton was found during excavation and intensive sieving (coll. Rabenstein Castle Museum – only bear bones were left in place).

Within this bone field a convincing *in situ* carcass situation is a vertebral column, which was found in two parts (Fig. **9**) being interrupted in the anterior thoracic area due to large carnivore scavenger impacts (see also Chapter 5). At the cervical vertebrae C2 to T1 found in full anatomic contact, even the first paired costae were found (Fig. **9B**). Below those a remarkable pathologic fibula was discovered (see Fig. 16). Several of the bones, *e.g.* one incomplete humerus without proximal joint (Fig. **9A**) and many other bones have bite damages (see details Chapter 5), such as the anterior thoracic vertebrae of the column (Fig. **9**).

All the longbones, such as the herein photographed nearly complete humerus, two ulnae, and one radius are in same small size proportions, as figured for the "Benno Skeleton" of the Bear' s Passage (Figs. **5 - 6**), whereas a sexual dimorphism analyses is not made yet on the few complete material. Using the few material of the Reindeer Hall bonebed, by the total length comparisons of the lower jaws and skulls, smaller females can be distinguished from larger males preliminary (Fig. **11**), which must be confirmed in future with more material.

Secure cave bear subspecies identification came from a complete skull which was also found in the bone field area (Fig. **10**) in the yellow sands below the speleothen layer (in a possible hibernation nest area = dark yellow marked). A thin speleothem layer is still strongly attached on the frontal. None of the complete lower jaws or two more mandible finds – all having differences in their length (= sexual dimorphism, or tooth use degree, Fig. **11**) fit in length, nor by their tooth polish stage to the skull. Another skull from the Reindeer Hall (private coll., Fig. **10A**) is also small and similar in shape, such as other *in situ* skulls, which are cemented in the sinter layer, especially those found under the large dropped ceiling block (Fig. **8**). Finally, also the Benno skeleton skull (Fig. **4**) is of this small cave bear type. In total, six skulls were mapped (two in the bone field, three under the block), a single heavy historically trampled and damaged skull (which braincase is nearly completely destroyed) was removed and rescued from the bone field with plaster.

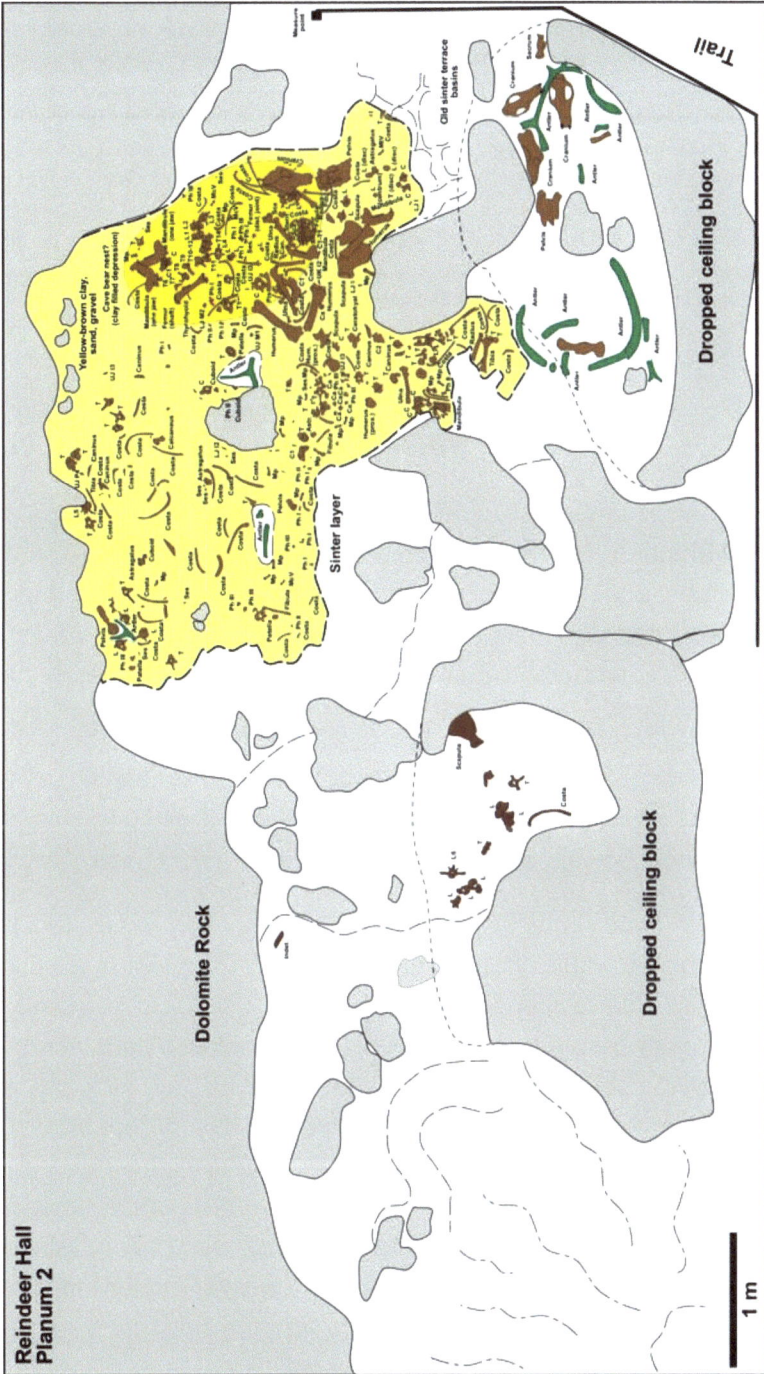

Figure 8. Bone field map of small cave bear *Ursus spelaeus* cf. *eremus* Rabeder *et al.* 2004 remains after the surface cleaning (and partly removal of heavy damaged sinter layer areas – due to skull or large bone removals in histioric times) in 2011. Below the mostly damaged final Late Pleistocene speleothem layer, from which the historically mentioned reindeer antlers, cave bear skulls, wolf and hyena skulls must have been taken, mainly by smashing the speleothem layer. In the next 20 cm of yellow-brown clays and sands many cave bear bones (brown) were excavated (today all in place). Within this "bone field" most of the remains represent many different animals from cub to adult sized animals. At least one nearly complete vertebral column is in place from a single individual. This has bite damage within the neck, on the first thoracic vertebrae, whereas a high amount of bones have large carnivore bite damage and bite impact marks. Even a few reindeer antlers have been found (green) with modern damaged fragments on the surface, and have been found in place only in one niche, and with several specimens below a large ceiling drop block, all within or directly below the speleothem layer, but not deeper. All those were encrusted and cemented by the sinter terraces.

Cave Bear Carcasses, Bones and Hibernation Nests in the Millionary Hall

As a result of the massive Post-LGM collapse of large ceiling blocks (see Chapter 8), half of Millionary Hall was covered (Fig. **12**). What is non-visible below (?cave bear nests, skeletons, bones) cannot be studied now, or in future. Further post-cave bear time sealing of surfaces within the final Ice Age speleogenesis and growth of the sinter terraces (1-3 cm thick sinter layers) cover nearly half of the hall (Fig. **12**).

Single skulls, bone and articulated skeleton parts are still visible in the Millionary Hall (between Large Millionary stalagmite and Reindeer Hall), but those are cemented (covered by) the speleothem layer.

There are *in situ* three skulls in the area of the connection to the Reindeer Hall, three skulls along the walls of the anterior Millionary Hall area in the distal sinter basins and an articulated thorax (cervical to middle thoracal vertebrae with rib cage, Fig. **12**). Individual cave bear bones (rib, metapods), and even in modern history, damaged rare wolf coprolites were found in the formerly reworked sediments in the area of the "damaged speleothem layer", opposite the Large Millionary Hall (marked in yellow in Fig. **12**). Similar as in the Reindeer Hall, the Millionary Hall has partly in place at least one cave bear carcass remain, which proves – cave bears hibernated and died there, and were not "washed into the cave". This is further supported by the new discoveries of the for long overlooked hibernation nests, which would either not have survived floods.

In the Millionary Hall only, about nine cave bear nests called *Ursalveolus carpathicus* Diedrich 2011 [13] seem to have survived some ten thousand years (Fig. **12**) [3], similar to those well-known with the round-oval depressions from several other European Caves including some of Upper Franconia caves [13 - 16]. The nests are scratched into the "Middle Pleistocene" yellow series sands and clays. Those nests can be found in three niches, in two cases along the cave wall, and even below one large ceiling drop block (Fig. **12**), but many more nests must have been present within the hall. Most of the nests (1-4 and 6) are about 60 x 80 cm, smaller ones (fit for cubs), only the fifth and those below the drop block are larger, with sizes of 150 x 180 cm (fit for adult bears). The different nest sizes indicate the presence and hibernation of cubs (Fig. **12**) [13]. In another cave in Europe, the Urşilor Cave, Romania, the skeletons of a possible mother cave bear and its one year old cub were found in their small and large nests close to each other on an extensive hibernation plateau deep in the cave system [13]. This is comparable at least in the nest arrangement and their sizes *e.g.* for the Millionary Hall (nests 3-5).

Most of the nests were first destroyed during the final Late Pleistocene, within the ceiling collapse (large blocks) and last two post- *U. s. eremus* period speleothem genesis times partly covering the nests with sinter terraces (Fig. **12**). Further and modern nest destruction from historic times (cavers, visitors) can be seen quite well on three of the depressions (nest no. 2, 5 and 9).

In the Sophie's Cave Millionary Hall, the bones and skulls have been scattered from the complete articulated skeletons lying in the nests. In most cases, bones were floating down (direction Reindeer Hall, Fig. **13**) away during the main speleothem genesis period at the end of the Late Pleistocene. The only articulated vertebral column of a cave bear skeleton, in the Millionary Hall, a second skeletal remain in the Reindeer Hall, and finally several articulated vertebral columns from the Bear's Passage, indicate that bears died during their "last sleep" (Fig. **13**), or were killed in their hibernation areas.

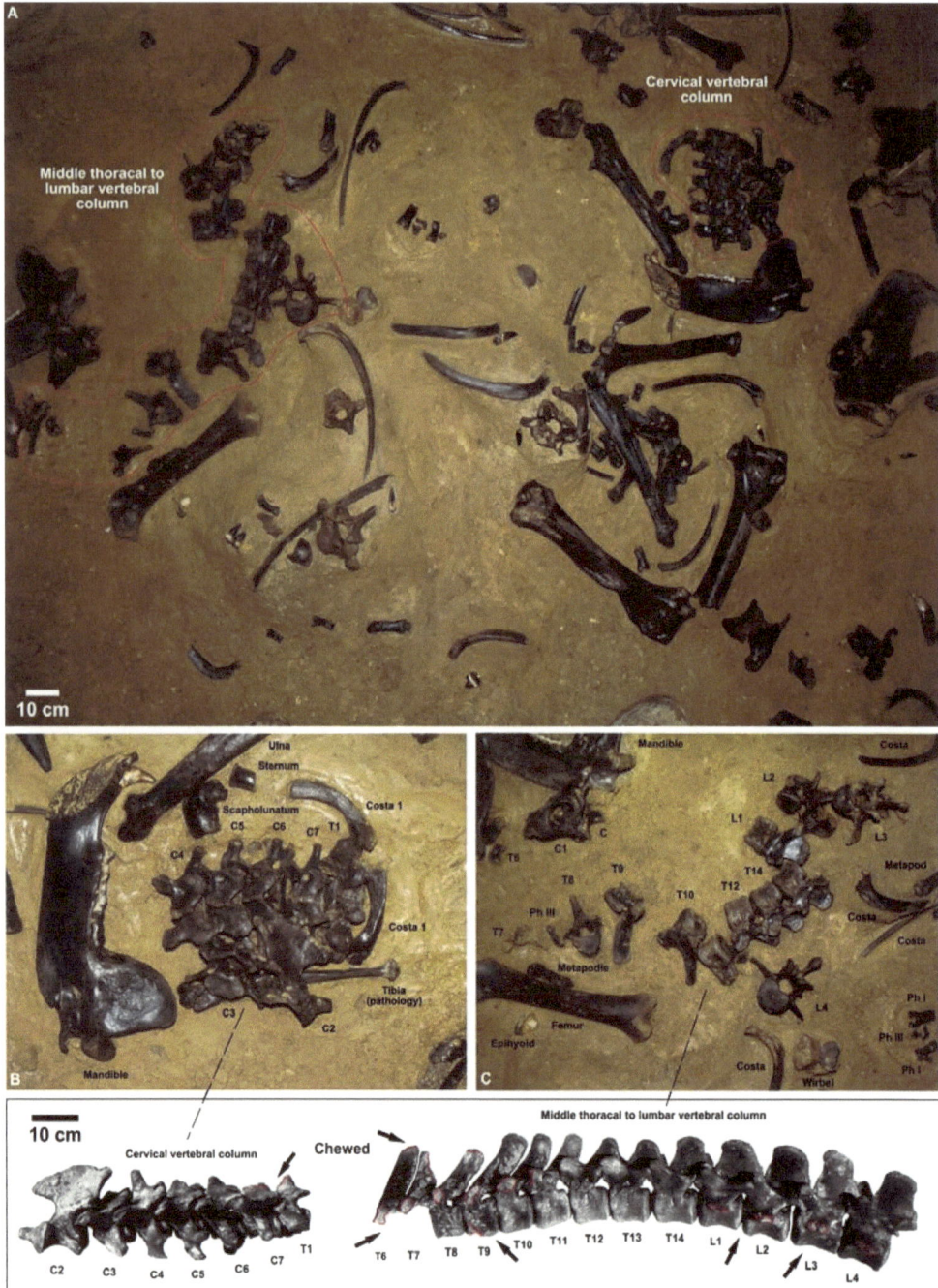

Figure 9. A. Bone field of the small cave bear *Ursus spelaeus* cf. *eremus* Rabeder *et al.* 2004 articulated remains within the Reindeer Hall, and **B-C.** Articulated, chewed vertebral column of an adult cave bear (seems to be a single individual) among the scattered bones of other individuals, including cub remains. This proves: cave bears hibernated also here in the lower part of the Reindeer Hall and were eaten, scavenged and skeletal remains were damaged and scattered by large predators, even deeper into the cave (coll. Rabenstein Castle Museum – all left in place in the Reindeer Hall bone field, cf. Fig. **5**).

Dating of the Cave Bear Population of the Bear's Passage, Reindeer Hall/Millionary Hall

In total, today, 13 cave bear skulls (in former reports about 30 were mentioned) from the small cave bear species are preserved in the Bear's Passage to Millionary Hall, plus a complete lower jaw, and three mandibles from each individual and different sexes which were recently excavated in the Reindeer Hall bone field (Figs. **4, 10, 11**). Those cranial remains with P4 teeth and several isolated P4 teeth from those same cave areas are the most important finds for dating of the older Sophie's Cave cave bear population.

After the comparisons of different cave bear species/subspecies skull shape morphology differences at the Harz Mountain Range cave bears from the Hermann's Cave and Baumann's Cave [12], those small skulls from the Bears's Passage, Reindeer Hall and Millionary Hall of the Sophie's Cave can be identified to belong to the small *U. s.* cf. *eremus* Rabeder *et al.* 2004 subspecies.

Their skulls have a variable saggital crest shape which is flat to slight convex (cf. Figs. **4, 10**) and are on average around 10 cm shorter as the largest *U. ingressus* skulls [12] (therein even stronger shape sizes due to sexual dimorphism), which saggital crest are also variable, but very long and nearly horizontal to slightly angled [12] (compare also Fig. **4** and Chapter 6 Fig. **4**, both plates in same scales). Those skulls have all a primitive three-coned P4 tooth enamel ornamentation (Figs. **10B, 11**).

This P4-morphotype distinguishing method [19 - 20] is important for the dating of the Sophie's Cave cave bear population (Fig. **14**) whereas this is calibrated herein also with newer European and Upper Franconia C^{14} dated *U. s. eremus* teeth [4 - 9] and the more recent distinguished skull shape forms [12]. All those three methods: 1. P4 morphotype, 2. Skull shape, and 3. DNA, are not yet well combined in an interdisciplinary study, that includes the newly rediscovered and since 1794 forgotten "cave bear skull" holotype of "*Ursus spelaeus* Rosenmüller 1794" from the Zoolithen Cave [13], which is still not DNA tested. It is recently believed to represent most probably after its skull shape and P4 morphology even the large "*Ursus ingressus* Rabeder *et al.* 2004", which latter then is an invalid younger synonym [13]. If this is correct, then the taxonomy of the cave bears have to be rewritten and renamed again in future for at least *U. s. spelaeus* and *U. ingressus*, and the "large cave bears" (see Chapter 6) would have to be revised to be spelled *Ursus spelaeus spelaeus*. Indeed, recently this name is used for another "haplo-type" [4 - 9], which makes it highly confusing in the taxonomic identification of European cave bears and subspecies. Most probable, the large cave bear "*U. ingressus*" species of the Sophie's Cave (Chapter 6) has to be renamed in the future.

The Sophie's Cave small cave bears from the Bear's Passage, Reindeer Hall and Millionary Hall range in general age, after the P4 method [20], and in comparison to other partly tooth C^{14} dated cave bear bones and sites [4 - 9], are between 113.00-32.000 BP of the early/middle Late Pleistocene to beginning of latest Late Pleistocene (Fig. **14**) [3]. Additionally, the other molar (upper and lower M1-3) tooth enamel surfaces with less amounts of cones (Figs. **4**, **10**, **11**) of the small cave bear subspecies fit to an older and more primitive cave bear dentition of not fully herbivorous cave bears [19], which is also known for other cave bear den sites in Europe [4, 9 - 11, 13 - 15, 17, 19].

On the relatively small canines of *U. s. eremus*, the sexual dimorphism is not well developed (cf. Figs. **4**, **10**, **11**). Canines in both, males and females, are only 20-21 mm in width (measured on the canine cusp enamel base), which can be seen also as plesiomorph character in early Late Pleistocene cave bear forms. This is much different in the younger large *U. ingressus*, which have a strong size difference in the canines. They are about twice in width in male canines, which allows identification of skulls and jaws in the sexes [9, 11, 12, 15, 17] (see Chapter 6). The small cave bear subspecies must be determined additionally in the future by DNA analyses (if possible), which is the only way to identify "haplo-types" and cave bear species/subspecies.

A

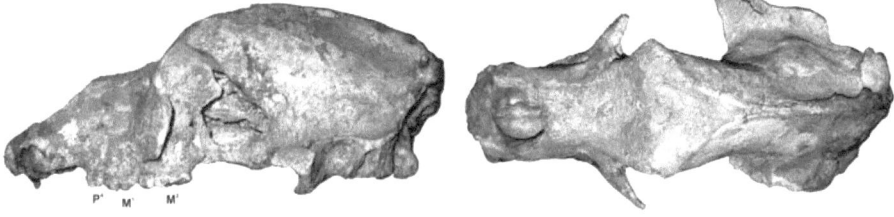

B

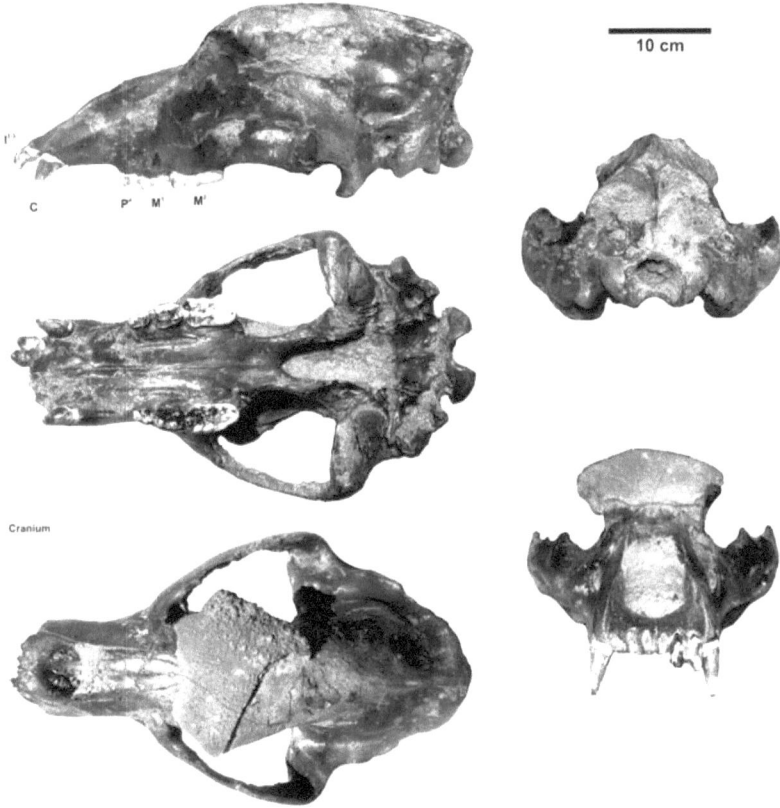

10 cm

I'³

C   P⁴ M¹ M²

Cranium

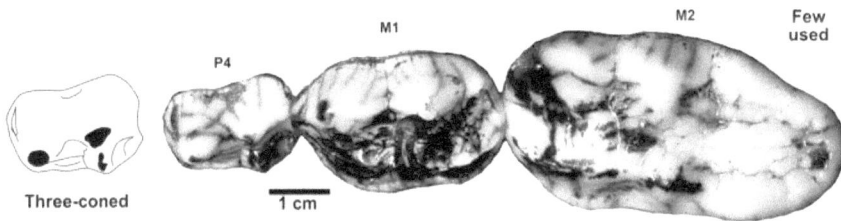

P4   M1   M2   Few
                                          used

Three-coned   1 cm

Figure 10. Small, early cave bear *Ursus spelaeus* cf. *eremus* Rabeder *et al*. 2004 skulls from the Reindeer Hall, Sophie's Cave. **A.** Older adult with medium used teeth and three-coned P4 (coll. Buchhaupt). **B.** Adult individual of best reproductive age from the bone field of the Reindeer Hall. The covering speleothem layer is still heavily attached to the frontal. The non-used dentition, especially the three-coned P4 teeth, also date the bones there into the early Late Pleistocene (coll. Rabenstein Castle Museum, *in situ*, cf. Fig. **11**).

The small cave bears from the Sophie's Cave are, however, not expected to represent the small alpine *U. s. ladinicus* types [4, 7, 8], but seem to belong to the well distributed middle mountainous boreal forest European small forms of *U. s. eremus* Rabeder *et al*. 2004 which date in most cases into the early/middle Late Pleistocene [3 - 9, 12 - 15, 17], whereas an overlap occurrence of both species *U. s. eremus* and *U. ingressus* and even brown bears is reported in the Swabian Karst region of southern Germany [6, 9], which cannot be confirmed in the Sophie's Cave.

Allthough there is a general "Late Pleistocene" age for the small cave bears in a restricted area of the Sophie's Cave (Fig. **1**), there are small differences within the Bear's Passage and deeper cave area small cave bear P4 morphologies (Fig. **14**).

Whereas the P4 tooth material of the Reindeer Hall is three-coned only, the P4 teeth from the Bear's Passage have already transitioned morphotypes ranging from three, to four coned tooth types, which date into the Early Late Pleistocene to Middle and even beginning of the Late Late Pleistocene [20] (Fig. **14**). It remains unclear, if the most primitive three-coned ones also are older and from the older Late Pleistocene Eemien Interglacial (120.000-113.000 BP), or even much older final Middle Pleistocene (late Saalian) age.

Such evolutionary stages between "deninger bears" and "early cave bears" are still problematic in the taxonomy and are often confused, but such P4 transitions are similarly found in the Zoolithen Cave Saalian (late Middle Pleistocene) to Weichselian (Late Pleistocene) fluvial mixed bonebeds [13].

The Bear's Passage might represent the most complete biostratigraphy of *U. s. eremus* and transitional *U. s. spelaeus* forms, but further material and excavations are of need from this cave branch. Stratigraphically few higher elevated finds (see Chapter 6) and multiple-coned P4 teeth of *U. ingresus* already date into the final

Middle to beginning of the late Pleistocene (approx. 32.000-24.000 BP, Fig. **14**) [4 - 9, 19, 20].

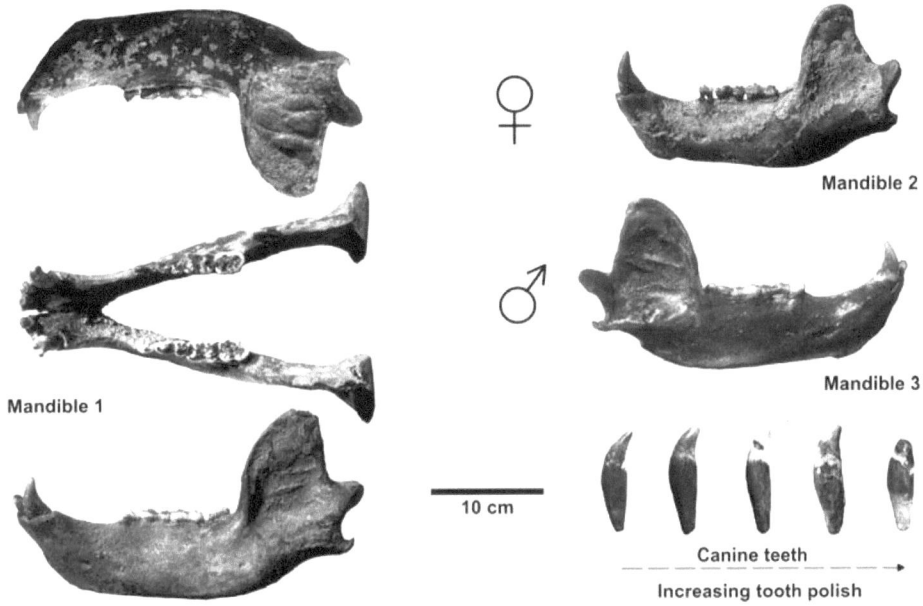

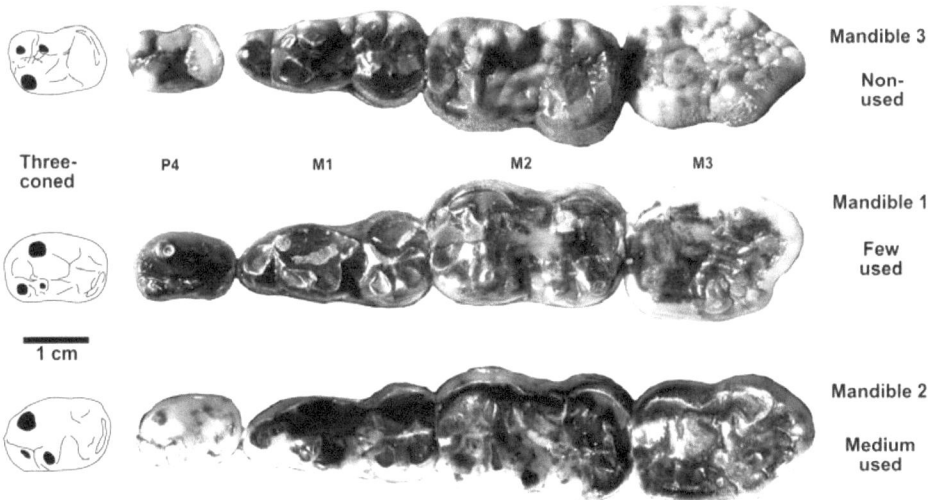

Figure 11. Small, early cave bear *Ursus spelaeus* cf. *eremus* Rabeder *et al.* 2004 lower jaws and mandibles of adult males and females at different ages (tooth use) from the bone field of the Reindeer Hall, Sophie's Cave (cf. Fig. 11). The three-coned P4 teeth date the layers and none of the material below the covering speleothem layer into the early Late Pleistocene. The sexual dimorphism is not easy to recognize on the polished canine teeth from these small, early cave bears (coll. Rabenstein Castle Museum, in place, cf. Fig. **5**).

The biostratigraphic dating of the older cave bear forms of the Sophie's Cave supports the theory of the blocking of the former entrance of the Bear's Passage, and a blocking event of the diagonal shaft between the Bear's Passage and Reindeer Hall, above the today's "Super Dieckmanns stalagmites" (Fig. **1**). There, only a small hole is left in the rocks today, which is partly closed by speleothems of the second (approx. 42.000 BP) and last (between 16.000-14.000 BP) generation.

This also means, the small cave bears penetrated first, starting in the ?Late Saalian/Eemian, but secure in the early Late Pleistocene, the Sophie's Cave Reindeer Hall and Millionary Halls over the Bear's Passage as deep as possible in the cave system for hibernation purposes.

Within the Middle Late Pleistocene (possibly correlating to second speleogenesis time around 42.000 BP, see Chapter 8), this became impossible, because a large block closed the small passage between Bear's Passage and Reindeer Hall, and only the Bear's Passage was in use by cave bears [3].

At that time, all large predators had easy access to the hibernating bears within a short distance of the cave entrance, which explains the heavily damaged cave bear bones there (see Chapter 5).

Somehow, at that time, the first ceiling collapsed during a humid period and the second speleothem genesis time also the Bear's Passage entrance must have been blocked, before the large *U. ingressus* species used the cave as a den (Fig. **14**) (see Chapter 6) [3].

Pathologies and Illnesses of Small Cave Bears

Only 0,1% of the 1.342 small cave bear subspecies *U. s. eremus* bones of the Sophie's Cave (Bear's Passage – most of the herein figured specimens), Reindeer

(Fig 12) contd.....

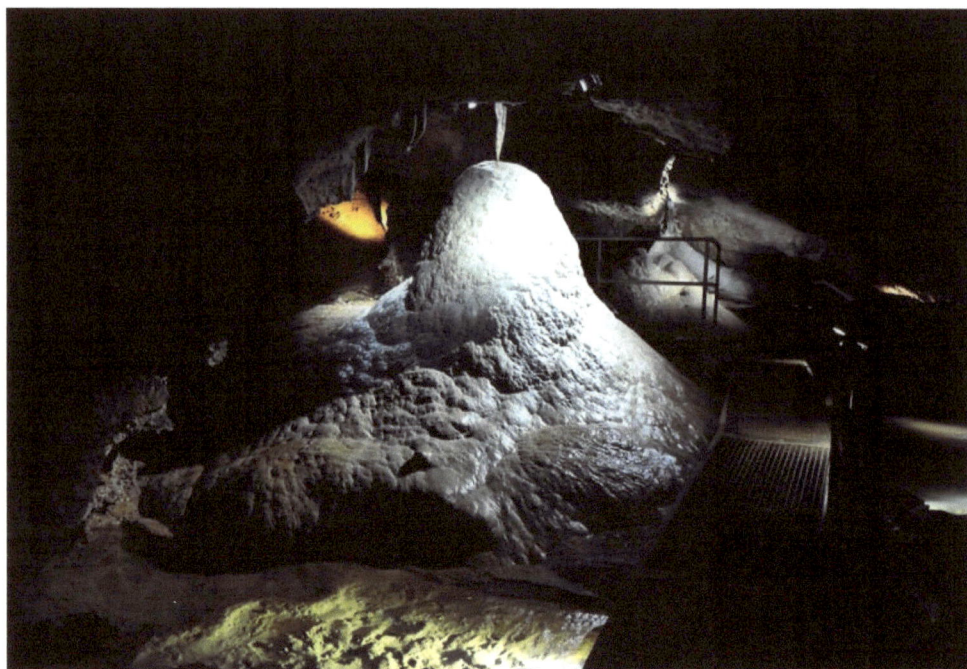

Figure 12. Above redrawing: Most probable preserved cave bear nests, which were found preserved in the corners and niches and even below large ceiling drop blocks covered in the Millionary Hall. Three cave bear skulls (two adults, one older cub) and an articulated adult cave bear were cemented in the final Late Pleistocene speleothem period by the sinter terraces, such as bones in the Reindeer Hall, Sophie's Cave. Below:. Photo of the Large Millionary.

Hall (only fibula) and Millionary Hall expose pathologies (Fig. **15**). There are two main groups of pathologies:

Not only in the Sophie's Cave, but also in other cave bear dens in Upper Franconia, mostly the Early/Middle Late Pleistocene cave bear types *U. s. eremus/spelaeus* Rabeder *et al.* 2004 lived within the Late Pleistocene glacial (Weichselian/Wuermian, 113.000-32.000 BP, Fig. **14**) [3 - 9, 11, 15] best known recently in Upper Franconia for the Zoolithen Cave [13], but also after first observations of new material of the Große Teufels Cave, and other nearby caves such as the Osterloch Cave or Hunas Cave ruine [14, 15].

Figure 13. Cave bears during hibernation (mother bear with one year old cub) in their nests within the Millionary Hall of the Sophie's Cave (Illustration G."Rinaldino" Teichmann).

Fractures are present on three middle thoracic ribs (possibly one adult individual due to similar colour preservation) which healing did not connect the ribs anymore. Also, one fibula did not heal completely distally. Furthermore, there are two on their proximal or distal articulation surfaces deformed phalanx II.

Exostoses (= bone growth) is found on several vertebrae, mainly on the anterior to middle thoracic bones. Exostoses bone growth in form of syndesmophytes is found at the vertebra centra such as the T2 and a middle thoracic vertebra. Isolated, not to the centrum attached syndesmophyte pieces have been found not connected to the bone itself. Exostoses are finally also recorded on the surface of a phalanx I.

Quite often "cave bears pathologies" were described/figured of other cave bear dens of Europe [21 - 24]. Older descriptions did not take the different smaller/larger species and subspecies or new knowledge about their predators (carnivores or humans) into account. Often, "cave bear" bone deformation or bone growth was in older descriptions misidentified as "arthritis, or osteoperoses" and was explained as a result of "cold-humid" cave climate [22 - 24].

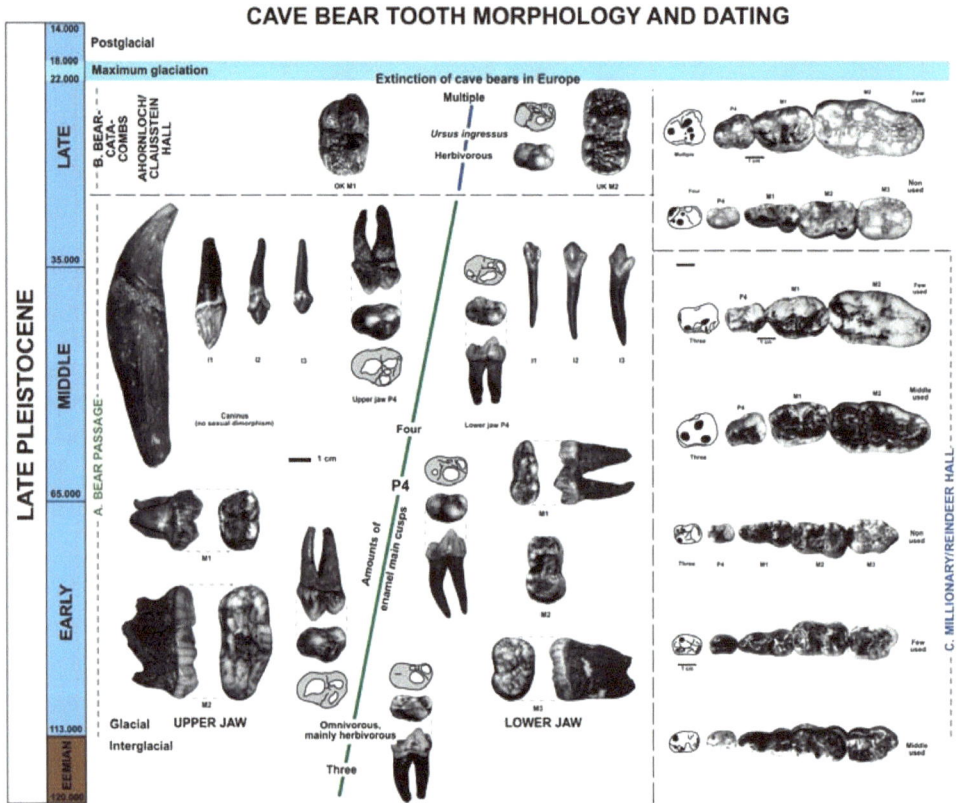

Figure 14. Dating of the small cave bears *Ursus spelaeus* cf. *eremus/spelaeus* Rabeder *et al.* 2004 and separation from the larger *U. ingressus* Rabeder *et al.* 2004 of different Sophie's Cave cave bear den areas. The few, coned, tooth enamel morphology date the smaller cave bears of the Millionary Hall into the Early Late Pleistocene, the ones from the Reindeer Hall and Bear's Passage into the Early to Middle Late Pleistocene, and large bears from the anterior cave parts (Bear Catacombs, Sand Chamber, Clausstein and Ahornloch Halls, Sophie's Cave) into the Late Late Pleistocene (some teeth are mirrored, from [3]). The small cave bears have no obvious sexual dimorphism within their canines, which is different in the large cave bears (all coll. Rabenstein Castle Museum).

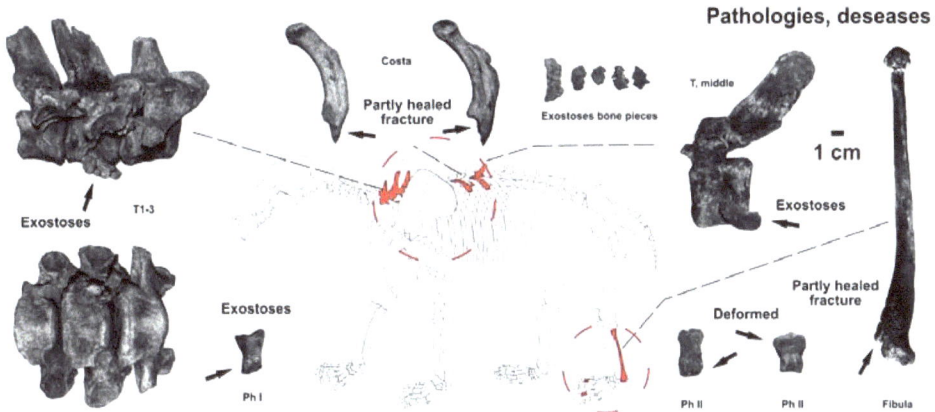

Figure 15. Pathologies and diseases on bones from small cave bear species *Ursus spelaeus* cf. *eremus* Rabeder *et al.* 2004, mainly from the Bear's Passage, and a few (fibula) from the Reindeer Hall, Sophie's Cave. Exostoses seems to be the result of fight injuries (with ?lions) which are mainly in the shoulder and the lower hind limb area (coll. Rabenstein Castle Museum).

Other misinterpretations on a skull from the Zoolithen Cave presented a cave bear skull with a hole in the frontal (in healing process) that was believed as "result of Neanderthal hunting ("spear or stone impact") [25]. This and a second skull from this cave, including many lion, hyena and wolf remains [13] are already revised to have resulted of "battles in the cave" between predating lions (or even hyenas), whereas fights between cave bears might also have caused such non-healed skull frontal bite damages [13, 21, 26, 27]. A revision of the "cave bear" pathologies is missing, especially including the new knowledge about "cave bear predation signs". The cave bear pathology origin in different cave bear species/subspecies includes now the overlooked predatory carnivore story. Bone pathologies on cave bear bones are, as demonstrated herein too, often a result of predatory stress [13, 21, 28, 29].

Fractures on the small cave bear subspecies bones of the Sophie's Cave might have happened during battles with other bears, but seem to be, more possibly, a result of lion and hyena attacks [21, 28]. Herein bite impact wounds and pathologies are found on the distal *U. s. eremus* hind limb (distal fibula, phalanx I). Recent Ice Age spotted hyenas research found, that the main trauma pathologies are found similar on the distal hind legs, because hyenas attack from behind even in interspecies battles, but also on prey [21, 30]. If the rib fractures are of inter- or intraspecies fights in origin, or if the bear fell or slipped in the cave on rocks, and thereby damaged its ribs, remains speculative. Exostoses pathologies can also originate from fights and bite or cut injuries on the bone "skin" on its surface [30], such as found at one phalanx I of the small cave bears.

Exostoses in the form of syndesmophytes are found typically on older "cave bear" animal vertebrae, and therein mainly in the neck and thoracic area, and more rare on lumbal vertebrae [12]. Such are found also on few anterior to middle thoracic vertebrae of the small cave bears of the Sophie's Cave (Fig. **15**). Those seem to attribute indeed to aged individuals only, which are also connected to osteoperosis, arthritis and the damages of the vertebrate centra cartilage due to aging [12].

REFERENCES

[1] Diedrich C. Ice Age geomorphological Ahorn Valley and Ailsbach River terrace evolution– and its importance for the cave use possibilities by cave bears, top predators (hyenas, wolves and lions) and humans (Late Magdalénians) in the Frankonia Karst – case studies in the Sophie's Cave near Kirchahorn, Bavaria. Quat Sci J 2013; 62(2): 162-74.

[2] Neischl A. Die Höhlen der fränkischen Schweiz und ihre Bedeutung für die Entstehung der dortigen Täler. Nürnberg, Schrag 1904; p. 95.

[3] Diedrich C. Late Ice Age wolves as cave bear scavengers in the Sophie's Cave of Germany – extinctions of cave bears as result of climate/habitat change and large carnivore predation stress in Europe. ISRN Zoology 2013; pp. 1-25.

[4] Hofreiter M. Genetic stability and replacement in late Pleistocene cave bear populations. Abh Karst-Höhlenk 2002; 34: 64-7.

[5] Rabeder G, Hofreiter M. Der neue Stammbaum der Höhlenbären. Die Höhle 2004; 55(1-4): 1-19.

[6] Stiller M, Baryshnikov G, Bocherens H, *et al*. Withering away - 25,000 years of genetic decline preceded cave bear extinction. Molec Biol Evol 2010; 27(5): 975-8.

[7] Rabeder G, Hofreiter M, Nagel D, Paabo S, Withalm G. Die neue Taxonomie der Höhlenbären. Abh Karst-Höhlenk 2002; 34: 68-9.

[8] Rabeder G, Hofreiter M, Nagel D, Whithalm G. New Taxa of Alpine Cave Bears (Ursidae, Carnivora). Cah sci Dép Rhône-Mus Lyon 2004; 2: 49-67.

[9] Bocherens H, Stiller M, Hobson KA, *et al*. Niche partitioning between two sympatric genetically distinct cave bears (*Ursus spelaeus* and *Ursus ingressus*) and brown bear (*Ursus arctos*) from Austria: isotopic evidence from fossil bones. Quat Int 2011; 245: 249-61.

[10] Post L. Building bear bones - a guide to preparing and assembling a bear skeleton or plantigrade possibilities. Lee Post 2004; 73.

[11] Diedrich C. Die oberpleistozäne Population von *Ursus spelaeus* Rosenmüller 1794 aus dem eiszeitlichen Fleckenhyänenhorst Perick-Höhlen von Hemer (Sauerland, NW Deutschland). Philippia 2006; 12(4): 275-346.

[12] Diedrich C. Evolution, Horste, Taphonomie und Prädatoren der Rübeländer Höhlenbären, Harz (Norddeutschland). Mitt Verb dt Höhlen- Karstf 2013; 59(1): 4-29.

[13] Diedrich C. Holotype skulls, stratigraphy, bone taphonomy and excavation history in the Zoolithen Cave and new theory about Esper's "great deluge". Quat Sci J 2014; 63(1): 78-98.

[14] Hilpert B, Kaulich B. Eiszeitliche Bären aus der Frankenalb - Neue Ergebnisse zu den Höhlenbären aus dem Osterloch in Hegendorf, der Petershöhle bei Velden und der Gentnerhöhle bei Weidlwang. Mitt Verb dt Höhlen-Karstf 2006; 52(4): 106-13.

[15] Hilpert B. Studies of the morphology of the bears from the Steinberg-Höhlenruine near Hunas. Abh Naturhist Ges Nürnberg 2006; 45: 117-24.

[16] Rabeder G, Nagel D, Pacher M. Der Höhlenbär. Thorbecke Species 4. Stuttgart, Thorbecke 2000; p. 111.

[17] Tsoukala E, Chatzopoulou K, Rabeder G, Pappa S, Nagel D, Withalm G. Paleontological and stratigraphical research in Loutra Ariedas Bear Cave (Almopia Speleopark, Pella, Macedonia, Greece). Sci Ann School Geol Arist Univ Thessaloniki (AUTH) 2006; spec vol 98: 41-67.

[18] Kaulich B, Rosendahl W. The Neonate cave bear skeleton from the Petershöhle near Velden (Franconia Alb, Germany). Abh Karst-Höhlenk 2002; 34: 12-6.

[19] Rabeder G. Neues vom Höhlenbären: Zur Morphologie der Backenzähne. Die Höhle 1983; 34(2): 67-85.

[20] Rabeder G. Die Evolution des Höhlenbärgebisses. Mitt Kommiss Quart Österr Akad Wissensch 1999; 11: 1-102.

[21] Diedrich C. The rediscovered cave bear "*Ursus spelaeus* Rosenmüller 1794" holotype of the Zoolithen Cave (Germany) from the historic Rosenmüller collection. Acta Carsol Slov 2009; 47(1): 25-32.

[22] Ehrenberg K. Der Höhlenbär. Naturw Monatsschr dt Lehr Naturk e.V. 1931; 44(1): 65-80.

[23] Moodie RL. La Paléopathologie des mammifères du Pléistocene. Biol Méd 1926; 24 (9): 431-40.

[24] Breuer R. Pathologisch-anatomische Befunde am Skelett des Höhlenbären. Spelaeogr Monog 1931; 7(8): 611-23.

[25] Groiss JT. Paläontologische Untersuchungen in der Zoolithenhöhle bei Burggeilenreuth. Ein vorläufiger Bericht. Erlanger Forsch B Naturwiss 1979; 5: 79-93.

[26] Diedrich C. The Late Pleistocene *Panthera leo spelaea* (Goldfuss 1810) skeletons from the Sloup and Srbsko Caves in Czech Republic (central Europe) and contribution to steppe lion cranial pathologies and postmortally damages as results of interspecies fights, hyena antagonism and cave bear attacks. Bull Geosci 2011; 86(4): 817-40.

[27] Rothschild BM, Diedrich C. Comparison of pathologies in the extinct Pleistocene Eurasian steppe lion *Pantherea leo spelaea* (Goldfuss 1810) to those in the modern lion, *Panthera leo* – Results of fights with hyenas, bears and lions and other ecological stress. Int J Paleopath 2012; 2: 187-98.

[28] Diedrich C. Cave bear killers and scavengers from the last ice age of central Europe: Feeding specializations in response to the absence of mammoth steppe fauna from mountainous regions. Quat Int 2011; 255: 59-78.

[29] Diedrich C. Extinctions of late ice age cave bears as a result of climate/habitat change and large carnivore lion/hyena/wolf predation stress in europe. ISRN Zoology 2013; pp. 1-25.

[30] Diedrich C. Inter-/intraspecies traumatic and dental/arthritic pathologies of cannibalistic Late Pleistocene spotted hyena *Crocuta crocuta spelaea* (Goldfuss, 1823) populations of Europe. (in review).

CHAPTER 5

LION, HYENA WOLF, WEASEL AND PORCUPINE CAVE DWELLERS - CAVE BEAR KILLERS AND SCAVENGERS

Abstract: In the early/middle Late Pleistocene, when the small cave bears *U. s. eremus/spelaeus* Rabeder *et al.* 2004 used a part of the Sophie's Cave as den, different large and small carnivors were mainly cave dwellers or short-time occupants. With the beginning of the Late Pleistocene glacial period (113.000 BP, terrace elevation 415 m a.s.l.), Ice Age wolves used one area in the cave as a den at the end of the Bear's Passage. *Canis lupus spelaeus* Goldfuss 1823 being represented only by grown up animal bones left larger amounts of phosphatic excrements in the cave bear bonebed especially in the Bear's Passage, but up to the Millionary Hall. A high percentage of about 26% of the cave bear bones have large predator bite damages. Mainly Ice Age wolves and Ice Age spotted hyenas scavenged the small cave bear subspecies carcasses. They produced larger bite damages on the vertebral column (inner side) proving an initial intestine/inner organ feeding. Steppe lions hunted cave bears even deeper in the cave where cave bears hibernated, whereas this can not be proven, only indirectly on large canine bite marks, which also might have resulted from those felids. Some cave area (also especially Bear's Passage) was used as a weasel *Mustela erminea* Linnaeus 1758 den, whereas the Ice Age porcupine *Hystrix* (*Acanthion*) *brachyura* Linnaeus 1758 dwelling is proven indirectly again only in the Bear's Passage with typical large rodent bite marks on two cave bear cub humeri. At the end of the middle Late Pleistocene, the former still unknown Bear's Passage entrance was blocked, which did not allow the smaller cave bears, carnivores or porcupines to penetrate the cave anymore, all inhabited/dwelled only a cave branch between approx. 113.000-32.000 BP.

Keywords: Late Pleistocene, sedimentology, Ailsbach Valley geomorphology, cave bear species/subspecies, cave bear clock, cave bear den, hibernation nests.

CARCASS AND BONE TAPHONOMY OF SMALL CAVE BEARS

On more than 26% of the small cave bear subspecies bones of grown up animals (subadult to senile) of the Bear's Passage, Reindeer Hall and Millionary Hall bite marks and bone damage (Figs. **1 - 3**) in different stages is present in similar

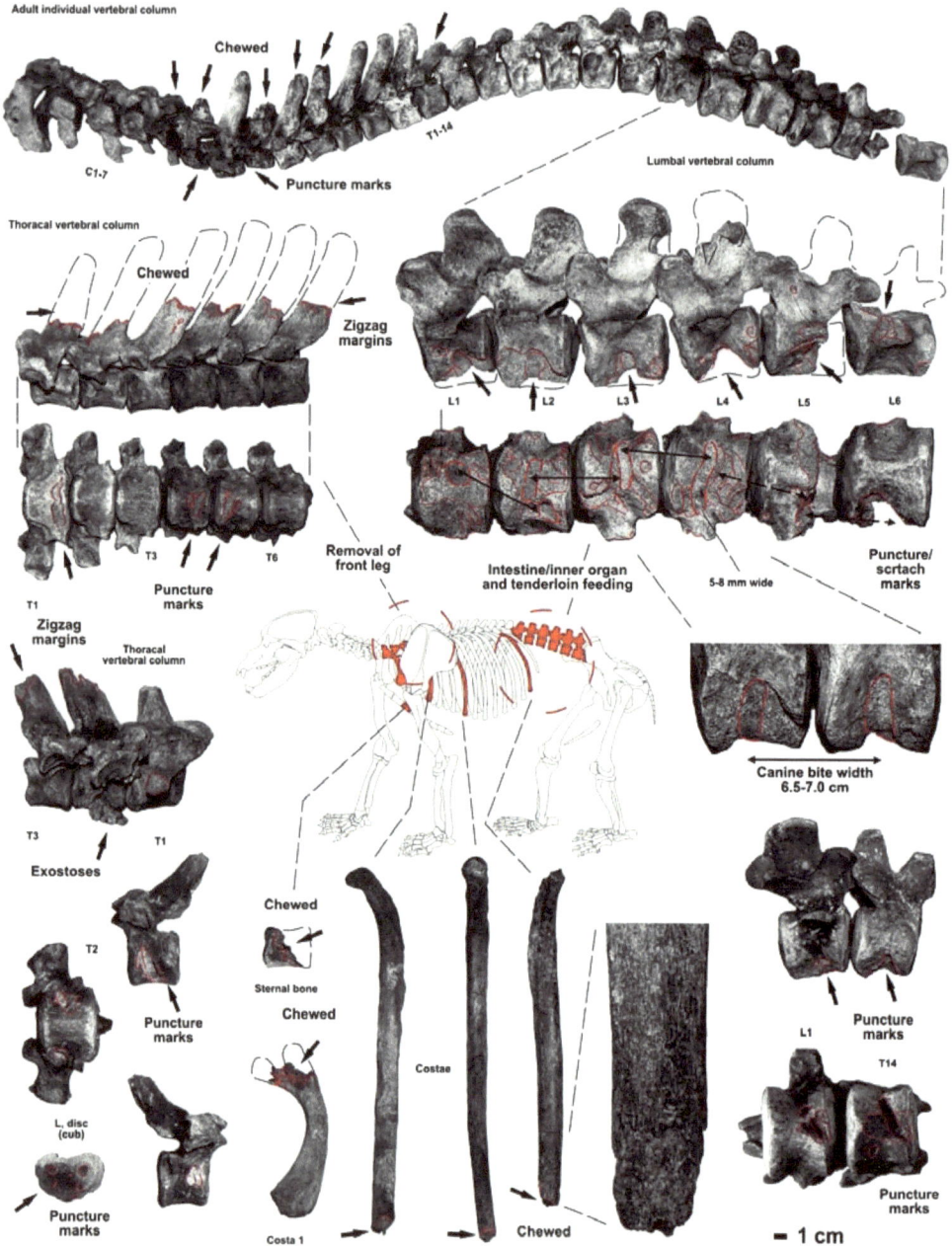

Figure 1. The large top predators (lions *P. l. spelaea*, hyenas *C. c. spelaea*, and wolves *C. l. spelaeus*) chewed and damaged small cave bear *U. s. eremus* vertebral columns and single bones from ther Bear's Passage, Sophie's Cave (coll. Rabenstein Castle Museum).

repeating forms. With high amounts of up to 60% of the small cave bear subspecies cub bones within the Bear's Passage and Reindeer Hall, those expose bite damage with puncture and bite marks in repeating similar forms, especially on the small longbones. First the soft distal joints and cartilage were eaten, whereas longbones were partly crushed to fragments.

The small cave bear carcasses were scavenged quite intensively, as remains are scattered (see Reindeer Hall bone field, Chapter 4, Fig. **8**), especially the skulls and legs, wheres all type of bones often have round-oval bite punctures, longbones were completely chewed on their distal joints; also vertebrae spines and costae expose chewing and stronger bite damages (Figs. **1 - 3**). The more or less articulated nearly complete cervical to lumbar vertebral column of a single small cave bear of the Reindeer Hall bone field has massive bite damage between the cervical and thoracic column (? shoulder blade region), and appears to originate from one carcass, that was initially decomposed by large top predators (see Chapter 4, Fig. **9**). If the bear had been killed there, or only the carcass had been scavenged, remains unresolved. The bones of at least six adult bear carcasses (estimated on the jaws and skulls) and possibly four older cubs between the age of six months and one year, show in many cases, bite impact marks or chewing damage on all kind of bones of different body regions (Figs. **1 - 3**).

Several articulated vertebral columns of adult cave bears (partially connected to ribs), from in total six partly articulated individual carcasses were documented from the Bear's Passage (about four lumbal vertebral columns, Fig. **1**), Reindeer Hall (one nearly complete vertebral column and some ribs in the bone field, cf. Chapter 4, Figs. **8 - 9**) and Millionary Hall (one anterior column with attached ribs within the sinter terrace, Chapter 4, Fig. **12**) prove their hibernation in those areas, and their death, naturally or caused by predators.

The vertebrae found in the Bear's Passage have deep, large bite, scratch, and puncture marks on the ventral sides in the centrum, being limited within the

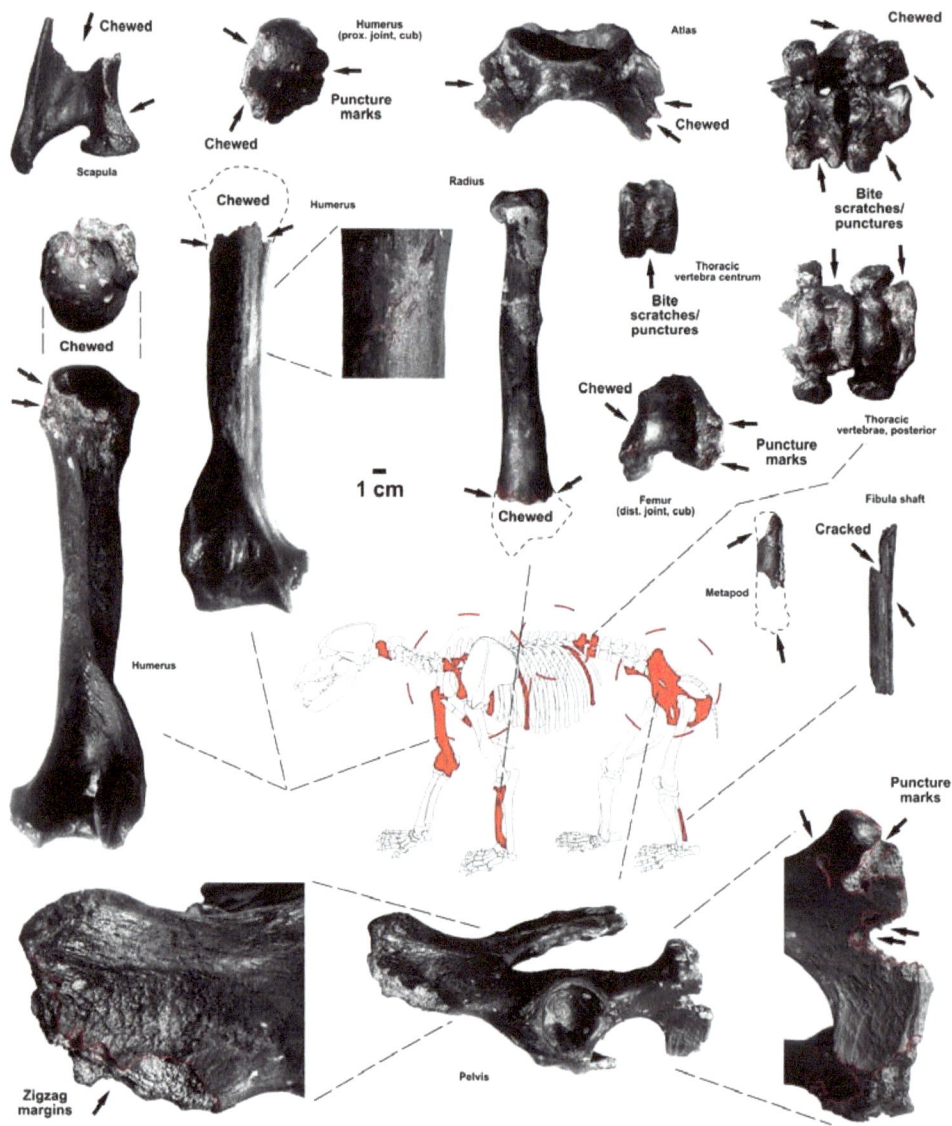

Figure 2. The large top predators (lions *P. l. spelaea*, hyenas *C. c. spelaea*, and wolves *C. l. spelaeus*) chewed and damaged small cave bear *U. s. eremus* vertebral columns, vertebrae, ribs, sternal bones and metapodials from the systematic excavated bone field in the Reindeer Hall, Sophie's Cave (coll. Rabenstein Castle Museum – all left in place in the Reindeer Hall bone field, cf. Chapter 4, Fig. **8**).

middle thoracic to lumbar region (Figs. **1** - **2**). The five to eight mm wide and up to 1.5 cm deep bite impact marks are round-oval depressions or scratches, which can be measured in their canine tooth scratch distances on at least one column (Fig. **2**).

The small cave bear material from the Bear's Passage to Millionary Hall allow to distinguish four different scavenged carcasses and bone destruction stages (Fig. **4**), and final bone chewing by porcupines. The main cave bear carcass/bone damage could have been produced only by the top predators: steppe lion *Panthera leo spelaea* (Goldfuss 1810), Ice Age spotted hyena *Crocuta crocuta spelaea* (Goldfuss 1823), and Ice Age wolf *Canis lupus spelaeus* (Goldfuss 1823) or in rare cases by the Ice Age leopard *Panthera pardus spelaea* (Bächler 1936) (Fig. **4**) [1 - 6]. The canine tooth scratch width is five to six cm and fit both: Ice Age spotted hyena or steppe lion skulls canine distances of the upper and lower jaws (Fig. **4**) [1 - 5]. Obviously, those impact bites into the cave bear body cavity were not made by the canines of the Ice Age wolf, which fang tooth distances are much lesser, also the canines are much thinner (Fig. **4**) [1 - 5]. Also leopards have much lesser canine tooth distances [6]. Wolves indeed, might have caused smaller round-oval canine tooth puncture marks 3 to 5 mm in width (Fig. **4**), but those seem to have been left only in softer porous bone parts (vertebrae, pelvis, Figs. **2** - **3**) [1 - 5].

Ice Age Steppe Lions as Cave Bear Killers and Soft Tissue Feeders

The large Ice Age steppe lions, which were similar in size (and even larger) to a small cave bear type *U. s. eremus*, must have penetrated the caves deep into the hibernation areas, where they hunted and killed cave bears [3, 7]. Most probably only or mainly cubs and/or weak grown up bears, because during hibernation they were an easy kill [3] this is best documented for the Romanian Urşilor Cave. Cubs were possibly mainly killed by lions (and possibly by leopards) within the caves [3, 6], buth were scavenged by all four top predators listed: *Panthera leo spelaea*, *Panthera pardus spelaea*, *Crocuta crocuta spelaea*, and *Canis lupus spelaeus* [1 - 12]. The final supporting proof for hunting in caves by steppe lions, which was based before on the bite marks on cave bear bones only, came by nitrogen isotope analyses, that could identify how cave bear cubs play a main prey role for steppe lions, especially at the end of the Late Pleistocene, when other game became

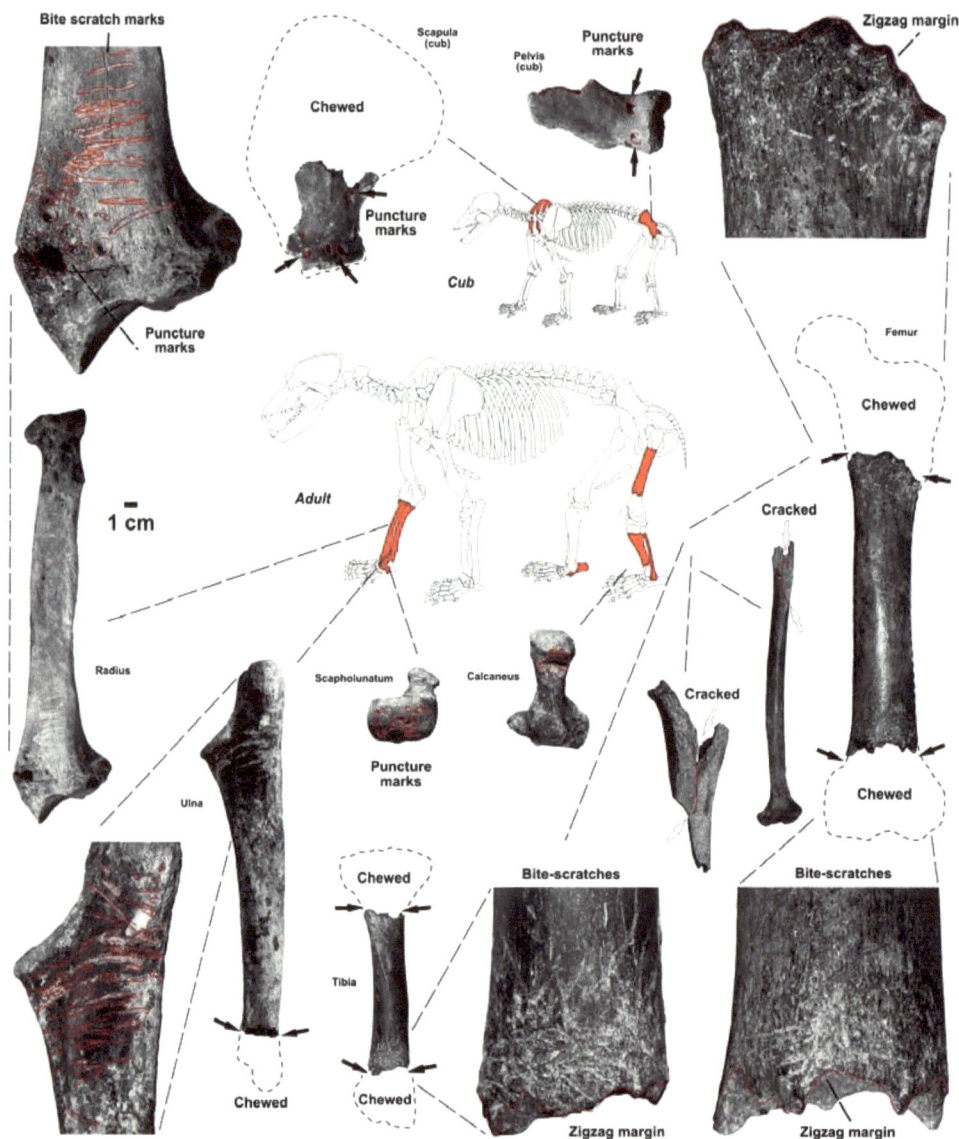

Figure 3. The large top predators (lions *P. l. spelaea*, hyenas *C. c. spelaea*, and wolves *C. l. spelaeus*) chewed, damaged and cracked (mainly by hyenas) small cave bear *U. s. eremus* bones from ther Bear's Passage, Sophie`s Cave (cf. Fig. **10**) (coll. Rabenstein Castle Museum).

scarce within the LGM (MIS 2) coldest period [13]. This cave bear hunt specialization is found only within boreal forest middle high mountain landscapes, as demonstrated additionally now in the Sophie's Cave, being already known for the nearby Upper Franconia Zoolithen Cave [4] or other caves in Europe [1 - 4, 7 - 12]. Therefore, cave bear cubs in "palaeopopulations" of cave bears do not have a "normal mortality" (= without predation impact) in all European cave bear dens where bone damage is always typically present in higher amounts. Normal death of siblings/cubs was suggested incorrectly by several non-carnivore studying cave bear researchers [14 - 19]. The small cave bears were also not consumed "cannibalistic by cave bears" as is formerly thought [18, 19].

In the Sophie's Cave [1], as well as in other cave bear dens in Europe [2, 9, 14 - 16, 20], there is also a high mortality rate in the *U. s. eremus/spelaeus* smaller cave bears in "reproductive" age, which cannot represent a "normal" mortality rate [21]. It seems more likely that those cave bears died, in several cases at least, in battles with the lions in the caves while trying to protect their young or themselves from being killed (see Chapter 6, Fig. **12**) [3, 4, 8 - 10, 21]. Of those attacked, weaker females could have been killed by a single lioness. Male/female ratio discussions on cave bear populations from European cave bear dens are also problematic and are not to use for the "classical understanding of their life and extinction, or cave use" [18], because the bone remains in the caves do not reflect 100% the real individual amount, nor sex ratio as we now know by the included predatory story [21].

If comparing the lions' jaw functions to other predators (leopards, hyenas, wolves) [6, 21] (Fig. **4**) the Ice Age steppe lions could have eaten (similar as leopards) only the softer parts, inner organs, intestines, meat and cartilage, and parts of the bone spongiosa of the joints (Fig. **4**) [21], similar in comparison to modern African lions [22, 23]. On several vertebrae and vertebral columns in the lumbar vertebra region on the ventral and lateral sides, 5-7 mm wide oval to elongated scratch and bite marks suggest scavenging into the body cavity by these largest predators, lions (or hyenas) [1 - 5]. Such bite marks and positions of lions/hyenas are similar to what was found on a Late Pleistocene Eemian Interglacial elephant carcass's vertebral columns [24] or on mammoth vertebrae [25]. The damage seems to have resulted less from "filet consuming" than "intestine/inner organ feeding", which lions generally eat first, which was monitored on a modern African elephant carcass [24]. Possibly, lions also left some chewed cave bear ribs or damaged cub bones in the Sophie's Cave, but did not completely "clean the carcass" after initial soft tissue feeding (Fig. **4**).

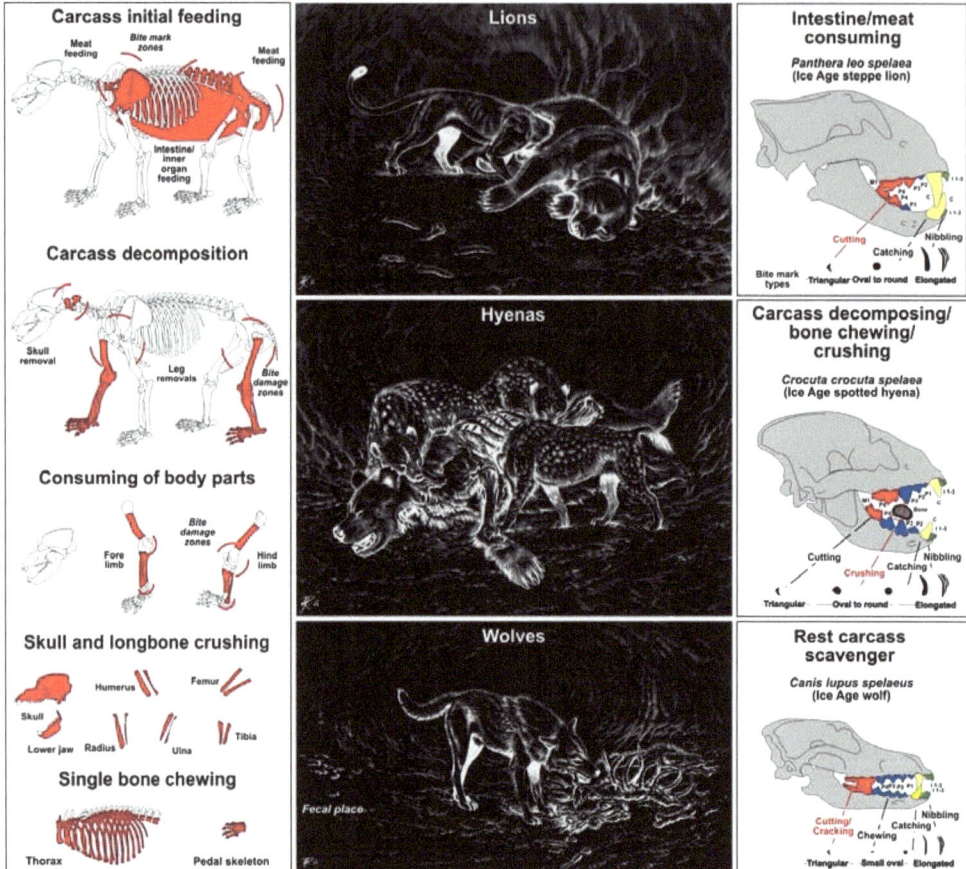

Figure 4. A picture story of the scavenging on cave bears by the three large Ice Age predators, and their differences in bite and jaw/tooth specializations for specific functions. **A.** Ice Age steppe lion on its cave bear kill, consuming only the intestines and inner organs, and possibly some meat (*e.g.* tenderloins on the inner lumbal vertebral column) using its meat-cutting dentition (carcass initial feeding, possibly initial carcass decomposition). **B.** Ice Age spotted hyenas destroying and damaging the cave bear carcass, including bone crushing with its bone crushing dentition (carcass initial feeding, decomposition, consuming of body parts and skull and bone crushing). **C.** Ice Age wolf consuming distal parts, soft ribs, and spongious parts of the pelvis and even the vertebrae, paws, and tail, that were left by other predators (single bone chewing). These wolves marked part of the Bear's Passage with their faeces, in which several cave bear bone fragments prove that they fed on cave bear carcasses and used the cave as their den ("cave imaging" illustrations by G. "Rinaldino" Teichmann; from [1]).

Ice Age Spotted Hyenas as Cave Bear Carcass Decomposers and Bone Crushers

The Ice Age spotted hyenas *C. c. spelaea* were bad climbers [21] similar to modern *C. c. crocuta* [26] and might have not killed any cave bears deeper within the caves, whereas this surely depended on the cave morphology: presence/absence of vertical/diagonal shafts/passages (see also cave model Chapter 6, Fig. **12**). A bear hunt by hyena clans to simply accessible cave parts must be expected, because even modern spotted hyenas have a keen sense of smell and see quite well in darkness [26]. Hyenas took any game prey carcasses apart in pieces, as a result of intraspecies competition and protection against lions (also leopard and wolves), which explains cave bear carcass scattering and also rare articulated cave bear skeletons in European caves [2, 3, 21, 27].

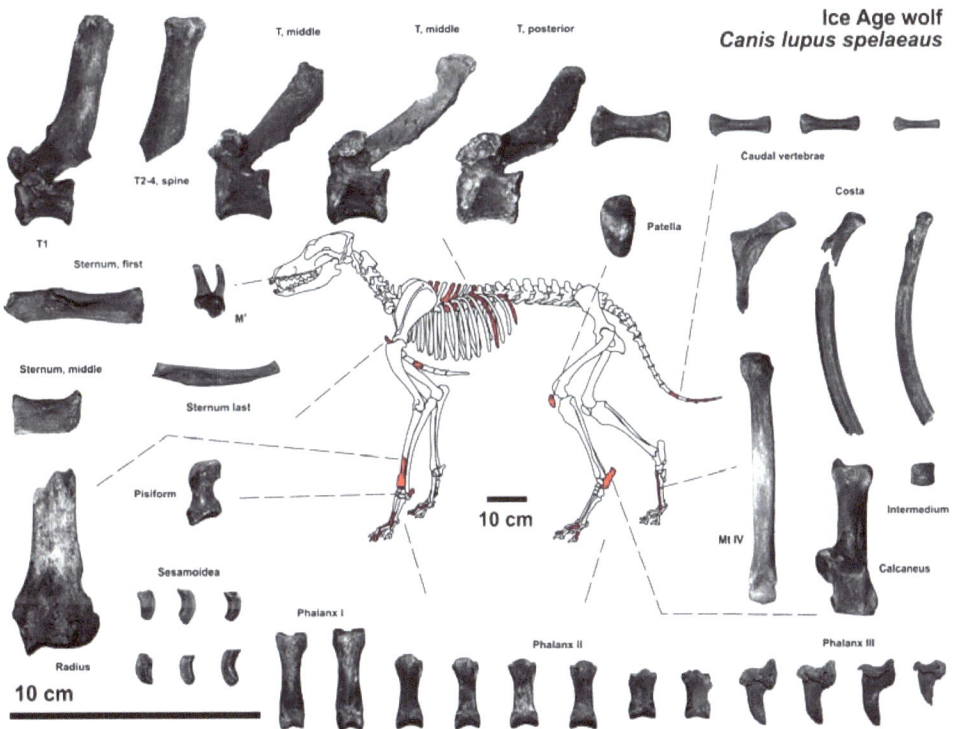

Figure 5. Remains of the Ice Age wolf *Canis lupus spelaeus* Goldfuss 1823 from the Bear's Passage, possibly of a single senile individual (coll. Rabenstein Castle Museum).

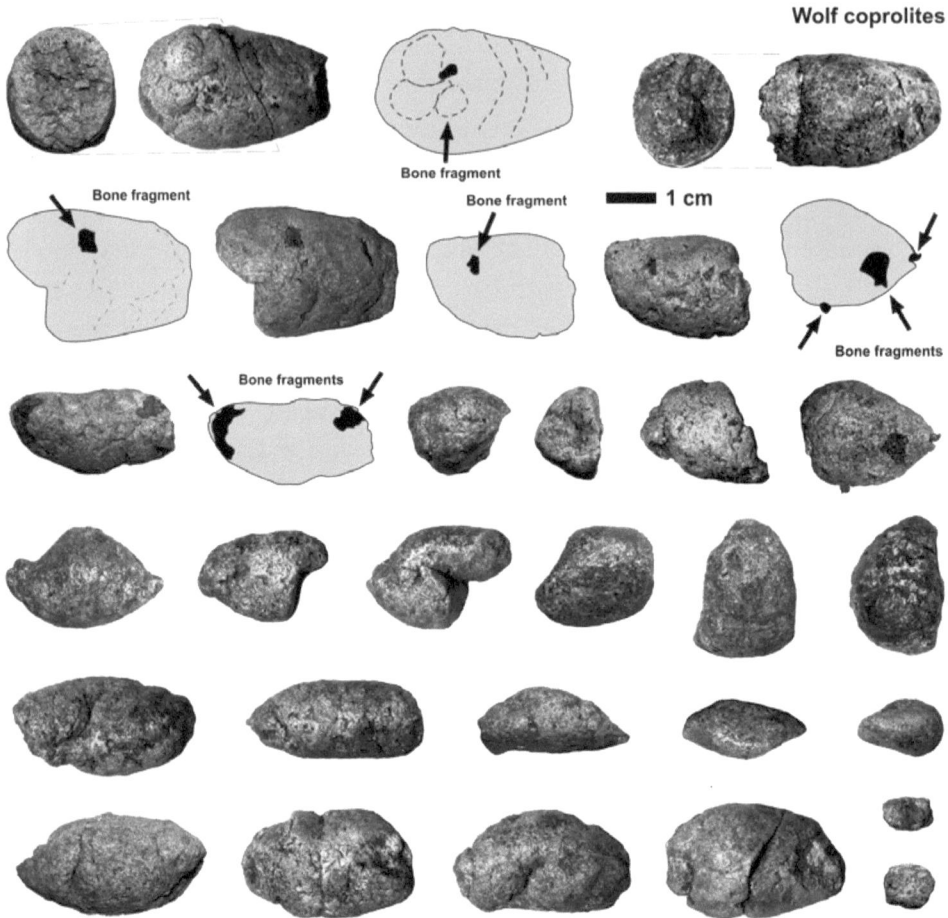

Figure 6. Ice Age wolf coprolites, partly including cave bear bone fragments from the Bear's Passage, Sophie's Cave (coll. Rabenstein Castle Museum).

The indirect proof of hyenas penetrating the cave bear dens is found less on "bite marks" on cave bear bones (which overlap with lion and wolf) such as more the abundant repeating similar bone three main damage stages (1. One chewed joint, 2. Bone shaft, 3. Bone fragment), which they also left on cave bear longbones [27, 28], similar as found herein on the longbones of the Sophie's Cave small cave bears (Figs. **1** and **3**). Well proven, also herein on the cave bear vertebral columns and massive chewed or crushed longbones (Figs. **1** - **3**) of the Sophie's Cave small cave bears, they must have scavenged what was killed in easy accessible areas by their own clans or was killed by lions/leopards [1 - 6]. As scavengers they surely simply must have consumed cave bears, that died of natural causes = bears not waking up after hibernation [2, 3, 5]. Spotted hyenas (few larger extinct *C. c. spelaea* and few smaller extant *C. c. crocuta*) are real bone crushers [21, 26] and were, in the Late Pleistocene with *C. c. spelaea* the main carcass decomposers and destructors of all mammoth steppe or mountain region boreal forest game in Europe even specializing on cave bears [2, 3, 21, 27] (Fig. **11**).

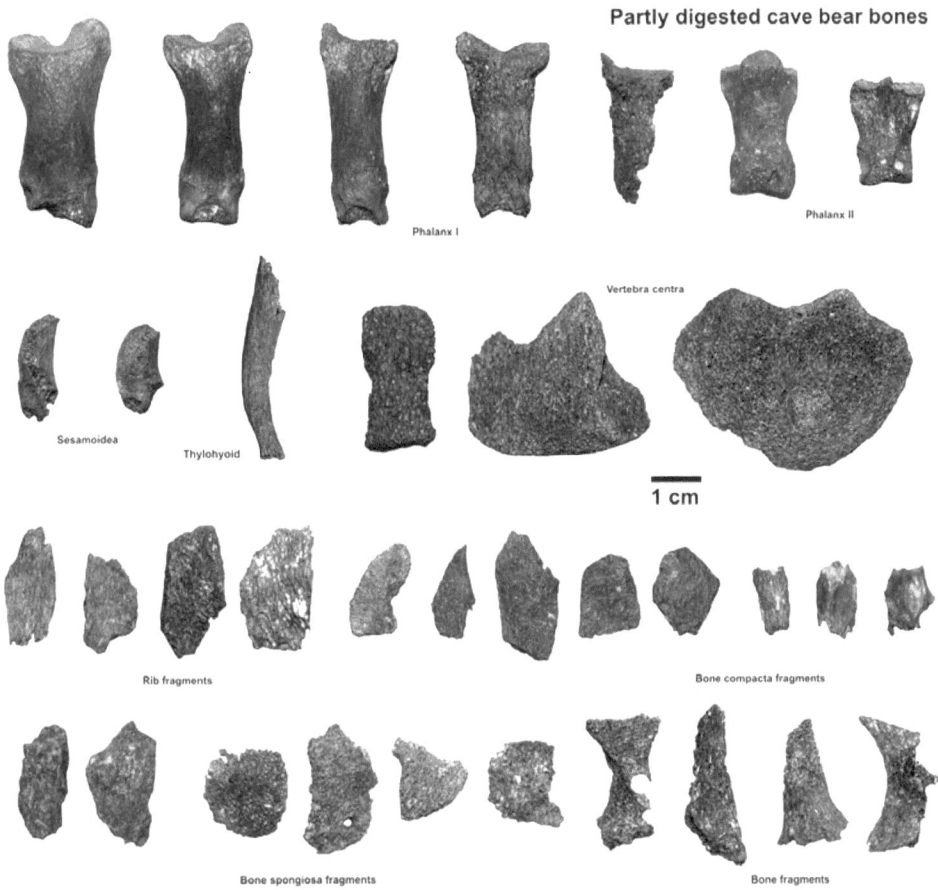

Figure 7. Bone fragments of different body regions which went through a wolf stomach, all from the Bear's Passage, Sophie's Cave. Although the surfaces are strongly corroded by stomach acid, it appears the carnivore ate the bear paws, indicated by many toe bones (coll. Rabenstein Castle Museum).

The cave bear prey longbones of grown up animals were finally crushed only/mainly by hyenas into pieces to reach the bone marrow, or to use the bone collagen of swallowed fragments [21]. This explains the presence of "thousands of cave bear bone fragments" and high percentages of damaged cave bear bones all over European cave bear den caves [2, 3, 21, 27]. Such fragments were recently also figured for the Upper Franconia Zoolithen Cave [10], which was even a hyena den in the entrance area, or the Hermann's Cave that was also a short-term used den [9].

The partly quick and seasonal or sporadic cave penetration by hyenas explains their own species bone record or complete absence in large cave bear den caves (*e.g.* Urşilor Cave [7], Hermann's Cave [9], Sophie's Cave herein), where those occupied only seasonally or short time entrance areas. Only within the den area, hyena teeth and bones are quite abundant [21, 29 - 36] (see also Chapter 6). From this period of the first cave bear den, a hyena den area was not found in the Sophie's Cave Bear's Passage, but might be expected within the undiscovered former and first entrance area.

The Wolf Den and Faecal Area in the Bear's Passage

The wolf den with a faecal place in the Bear's Passage is a unique European documentation of carnivores and their cave use. Only in the distal part of the Bear's Passage 52 wolf bones including a single upper jaw M^2 tooth were discovered new in 2010 (Fig. **5**). Those belong most probably to a single individual of an adult wolf, which has pathological damage at the middle thoracic vertebra dorsal spine (Fig. **5**).

It was a large Ice Age wolf subspecies *Canis lupus spelaeus* Goldfuss 1823 [10, 37] which was smaller compared to the Canadian arctic-boreal mountain adapted timber wolf [38] and was larger as the modern European wolf that has not yet been well defined by DNA [39]. Wolf palaeopopulations are also known from other caves in the area (the Zoolithen Cave [10, 38] and Große Teufels Cave after own observations).

Some postcranial bones have been compared, having similarly large proportions to those from the Zoolithen Cave, Sophie's Cave and Große Teufels Cave [38] where the skull and longbone sizes are between European and Scandinavian Arctic wolf and Canadian Columbian wolf subspecies [40]. They possibly belong to a specialized Late Pleistocene wolf ecomorph, which also became extinct before the LGM [39].

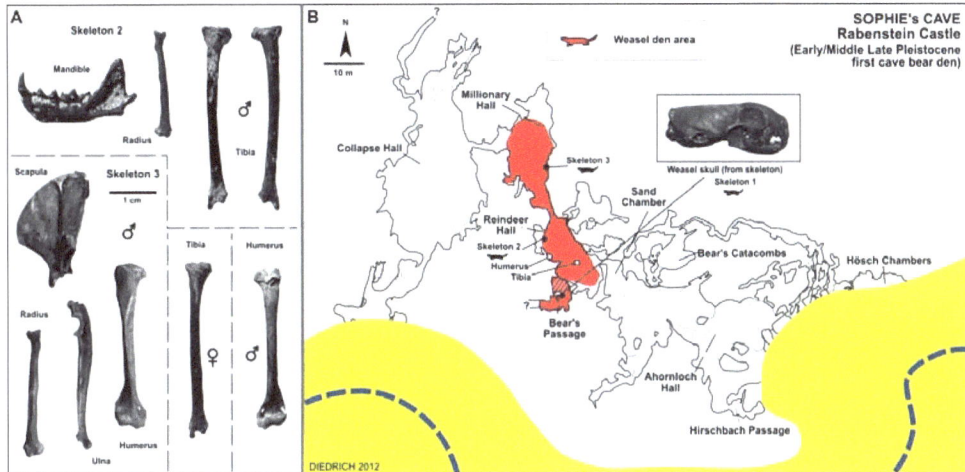

Figure 8. Overview of the four skeleton remains and single bones of the weasel *Mustela erminea* Linnaeus 1758 subsp. from the weasel den in the Bear's Passage up to Millionary Hall, Sophie's Cave.

Fossil Ice Age wolf *C. l. spelaeus* excrements were found within the Bear's Passage, Reindeer and even Millionary Halls [1]. Those most abundant from the Bear's Passage are found beside the wolf bones (see Chapter 6, Fig. **7**). The pellets from the Bear's Passage count 91 nearly complete and 160 fragments of about max. 4 cm elongated and medium-sized, grey phosphatic fossil excrements, so-called coprolites (Fig. **6**). They were sieved in 2010 mainly from the historically reworked sediments at the end of the Passage (see Chapter 6, Fig. **7**). Because of the large quantities of pellets and comparisons with the faeces of modern wolves [40 - 42], a reasonably firm attribution to wolves was made [1]. These coprolites are different in their shapes to much larger and different pellet aggregates of the Ice Age spotted hyena [43], which latter are only abundant in hyena dens due to their intensive den marking activities in the Late Pleistocene [21]. In several small fragments their last prey bones are cemented in the coprolite, which seem to originate from cave bears [1].

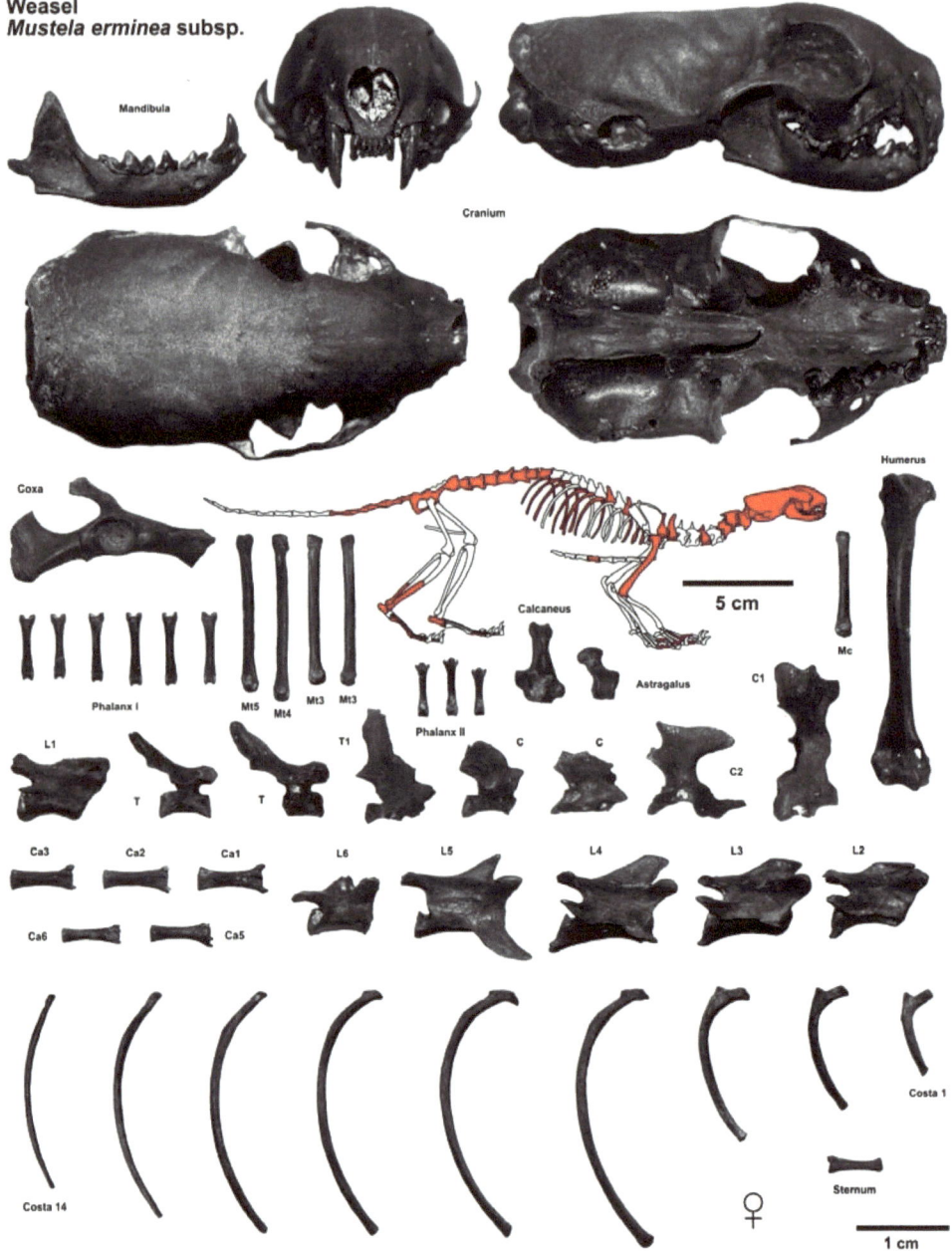

Figure 9. Remains of a single skeleton of the weasel *Mustela erminea* Linnaeus 1758 subsp. (female) found in a den burrow of the Bear's Passage, Sophie's Cave (cf. Fig. 20B, coll. Rabenstein Castle Museum).

Besides the coprolites at the end of the Bear's Passage, abundant "strange-looking" cave bear bones and bone fragments were extracted from the sediments during sieving. "Complete bones", *e.g.* phalanx I and II bones (Fig. **7**) are heavily corroded on their surfaces, which could have happened only by stomach acid [1]. Also "holes" in the bones (Fig. **7**) are typical stomach acid corrosion signs of carnivores [21]. Small fragments of bone spongiosa and compacta look quite similar (Fig. **7**). Those cave bear bones have been identified as having passed through "wolf stomachs" [1]. This is also indicated, with similar bone fragments sticking in the wolf excrements (Fig. **6**), which were examined only by microscopic work, but not with DNA studies, yet. Analyzing the cave bear bone types that went through a "carnivore" stomach, those are dominated by rib fragments, a few vertebrae spongiosa pieces, and most interestingly, of completely swallowed distal paw elements (phalangae, sesamoidea, Fig. **7**). In the Bear's Passage, it was added in the carnivore-bear conflict that wolves can be identified as the third large cave bear predator, but possibly, those were at least here at the Sophie's Cave only the final scavengers, after lions killed, and hyenas decomposed the carcasses (Fig. **4**) [1]. This is not unexpected, because even modern North American Timber wolves at least scavenge on black bear carcasses [44].

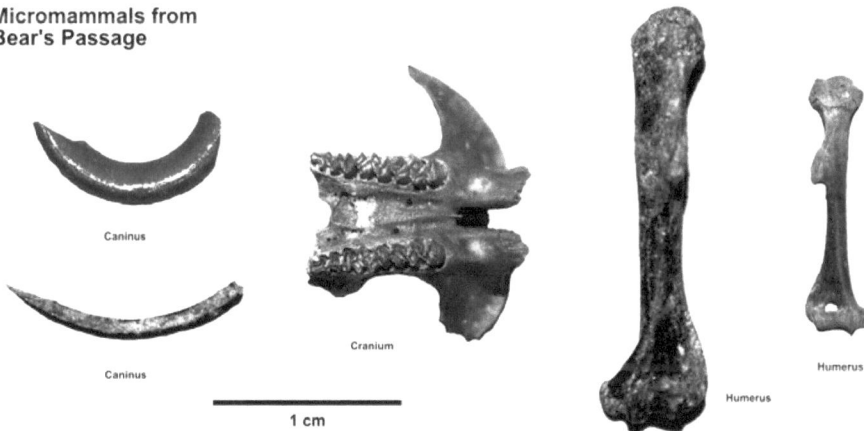

Figure 10. Late Pleistocene undetermined micromammal remains (canines, skull and humeri), most probably prey remains of the weasels from the Bear's Passage, Sophie's Cave (coll. Rabenstein Castle Museum).

WEASEL SKELETON IN THE BEAR'S PASSAGE – AN ICE AGE WEASEL DEN WITHIN THE SOPHIE'S CAVE

The Late Pleistocene weasel *Mustela erminea* Linnaeus 1758 is probably an extinct subspecies which is DNA untested [45 - 47]. The first most complete Late Pleistocene skeleton is another unique European discovery in the Bear's Passage of the Sophie's Cave (Fig. **8**). Furthermore, it is the first and exact taphonomic study of a Late Pleistocene weasel cave den. Even modern caves are not yet studied concerning their weasel den use and their accumulated prey remains.

The weasel skeleton of the Sophie's Cave has a complete skull with jaw including all teeth, most of the vertebral column and ribs, or sternal bones (Fig. **9**). The extremity bones (one humerus, half tibia/fibula), include many small pedal bones (16 phalangae, four metapods) because of sieving of the skull surrounding sediments (1 mm sieve). All remains were found in the Bear's Passage in an obviously refilled about 15 cm thick burrow (see Chapter 4, Fig. **7**). This is interpreted as the weasel den burrow, where the animal finally died, having its bones protected against scattering or trampling by scavengers or cave bears.

A second "skeletal remain" consisting of at least one mandible, radius, and both tibiae, both half humeri, a costa remain, and two vertebrae, was discovered within the bone field about 20 cm below the speleothem layer (Fig. **8**) opposite the Large Millionary stalagmite in the Millionary Hall close to the wall, and was half-way encrusted by thin slepeothem layer. A single *Mustela*-tibia was found during the extraction of the mammoth pelvic coxa (Fig. **8**). This single bone could even fit with the skeleton of the Bear's Passage. Possibly the tibia was transported from there to the middle part of the Reindeer Hall, which remains speculative. Another isolated bone, a humerus, was found from another individual in a niche in the lower part of the Reindeer Hall (Fig. **8**), outside the bone field.

After comparison of longbones of at least four individual skeletons or skeletal parts within the Bear's Passage, Reindeer Hall and the Millionary Hall (Fig. **8**), the most complete from the Bear's Passage fits to a smaller female, indicated best by its smaller humerus, ulna, radius and tibia proportions compared to the well-known sexual dimorphism in the modern weasel [45 - 50]. Comparing the longbones from the Sophie's Cave to finds from Austrian and other Upper Franconia caves [48, 51], they allow the identification of the larger proportioned longbones (single humerus) to originate from male weasels. The bone proportions of the longbones of the second skull-less skeleton fit best to a male. In total, one female and two males seem to be represented at minimum in the early/middle Late Pleistocene (as recently dated by cave bear teeth, see Chapter 4, Fig. **14**) "weasel population" with more finds expected with future sieving of sediments in the Sophie's Cave.

Figure 11. Late Pleistocene weasel *Mustela erminea* Linnaeus 1758 subsp. importing a mouse into its den area of the Bear's Passage, Sophie's Cave (Illustration G. "Rinaldino" Teichmann).

Table 1. Early/Middle Late Pleistocene megafauna bone assemblage (NISP = 1.463) from the Bear's Passage/Reindeer/Millionary halls of the *U. spelaeus eremus/spelaeus* bonebeds.

Species	Bone amount (= NISP)
Ursus spelaeus eremus/spelaeus	1.370
Canis lupus spelaeus	41
Mustela erminea	52

As is known for modern weasels, they use natural caves and burrows as dens [52, 53], whereas a burrow is present even within a cave (Bear's Passage, Chapter 4, Fig. 7). The Sophie's Cave was used simultaniously by small cave bears and weasels only in the Bear's Passage to Millionary Hall. Only one European site, an Austrian cave, has a similar abundance of mustelid remains, but all without find mapping or taphonomic context, being represented only by selected larger bones and no skulls [48] which seem to originate also from a weasel den cave. Other Late Pleistocene bone sites with *M. erminea* remains in Upper Franconia are the Zoolithen Cave and two other caves [48, 54], where at least four caves seem to have been weasel dens during the Late Pleistocene in Upper Franconia.

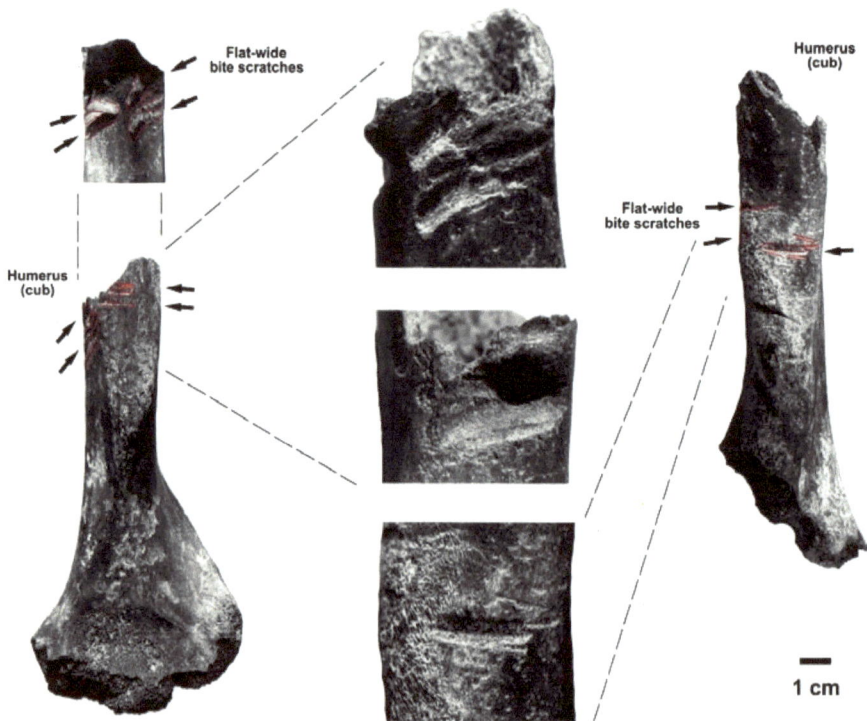

Figure 12. Late Pleistocene porcupine *Hystrix* (*Acanthion*) *brachyura* Linnaeus 1758 canine bite scratches: rectangular and 4-5 mm wide bite marks, partly parallel left by both incisors, on two cave bear cub humeri from the Bear's Passage, Sophie's Cave (coll. Rabenstein Castle Museum).

Weasels can penetrate, as with martens, very deep into caves [52, 53], but the ones of the Sophie's Cave seem to have explored the cave no further than the "cave bear den area" most deep into the cave system as into the Millionary Hall (Fig. **8**). In caves, modern weasels can orient quite well as nocturnal hunters [52, 53]. They hunt mainly micromammals [52, 53] and obviously the small amount of Ice Age "mice bones, and a skull" (Fig. **10**, dark in colour similar to weasel and cave bear bones), also sieved from the Bear's Passage weasel den area, might have been imported, most probably by weasels as prey (Fig. **11**). The presence of such micromammals, which have not been found within the Reindeer Hall bone field bonebed yet, also underline the weasel den and use of this cave branch by a small mustelid carnivore.

The fur colour changing weasel (winter: white, summer: brown, and always black-tipped tail) is still today an arctic to temperate climate small carnivore [52, 53] living in the Pyrenees, Alps and Carpathians Mountains. It has also been found in the north of Eurasia, Central Asia including Japan, Greenland, and Northern America and Canada [52, 53] and seem to have been, during the Late Pleistocene, a typical boreal forest inhabitant, especially in karstic regions such as Upper Franconia.

THE LAST PORCUPINES OF EUROPE

On two cave bear cub humeri from the Bear's Passage, very typical around 5 mm wide-non-deep bite mark scratches (Fig. **12**) from "large rodents", obviously porcupines [55] give indirect proof of their dwelling within the Bear's Passage (Fig. **13**). Those porcupines are the last ones in Europe with the species *Hystrix* (*Acanthion*) *brachyura* Linnaeus 1758 [55].

In Upper Franconia, such Late Pleistocene porcupines are not only proven indirectly by the chew damaged megafauna bones for the Sophie's Cave. Their own species teeth, lower jaws or few postcranial finds such as bite damaged prey bones (horse metapodial, bison metapodial, cave bear mandible) are reported and figured from Upper Franconia for three other smaller cave cavities, such as the the Zwergenloch Cave near Pottenstein, Hasenloch Cave, and the Fuchsloch Cave in the nearby surroundings [55 - 58].

Figure 13. Late Pleistocene porcupine *Hystrix* (*Acanthion*) *brachyura* Linnaeus 1758 chewing on cave bear cub longbones in the Bear's Passage, Sophie's Cave (Illustration G. "Rinaldino" Teichmann).

All over Europe, their sporadic presence or few tooth/bone records and chewed prey bones overlap often with Ice Age spotted hyena *Crocuta crocuta spelaea* (Goldfuss 1823) cave dens, where *Hystrix* reused hyena imported prey bones or carcass remains [21, 55]. In other cases, such as the Hasenloch or Zwergloch, the caves are very small and were possibly used only by porcupines, similar as known for modern African porcupine species, which dens also often overlap with different hyena species den cave sites, especially those of spotted and striped hyenas [26, 59].

It remains unclear, if the last porcupines of Central Europe migrated from Central Asia with the megafauna in the Late Pleistocene Eemian Interglacial, or the beginning of the Weichselian/Wuermian Glacial [55]. This biostratigraphic difference is important for the dating of the bonebeds (and cave bears) of the Bear's Passage in the Sophie's Cave, which might reach older (not only MIS 5d) into the Eemian Interglacial (MIS 5e) (cf. Chapter 4, Fig. **14**).

THE EARLY/MIDDLE LATE PLEISTOCENE (MIS 5D-3) BOREAL FOREST PREDATORS AND GUILD

The comparison of five main different bone assemblages in hyena den and cave

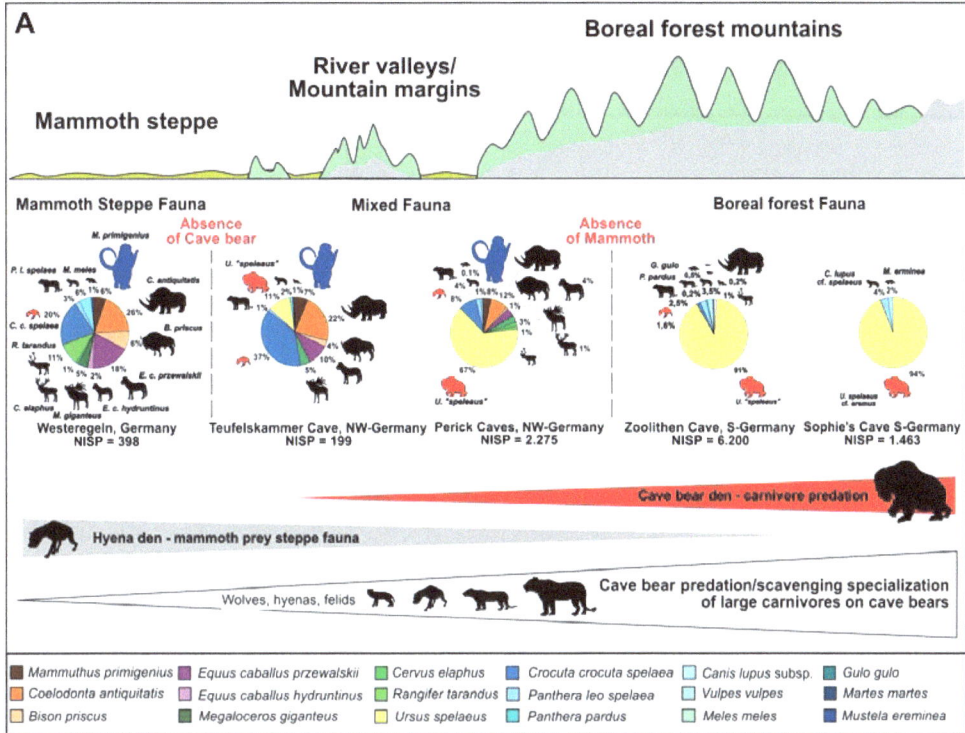

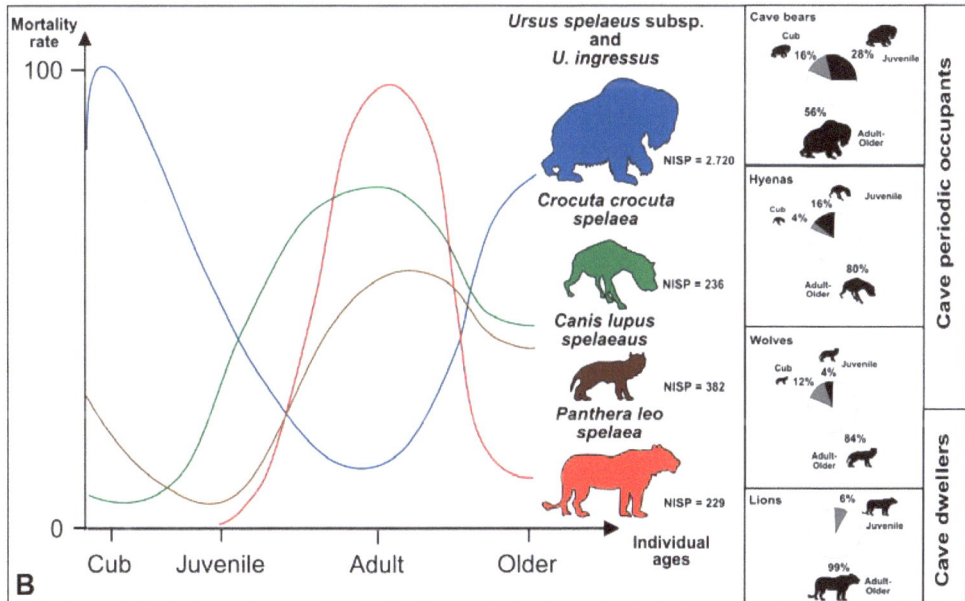

Figure 14. A. Comparison of three main bone assemblage types from the mammoth steppe to those from mountainous Boreal forest Ice Age palaeoenvironments in northern and southern Germany composed. Upper Franconia larger caves (Fig. 1) seem to contain only the "Boreal Forest Fauna" bone assemblage such as analysed well for the Zoolithen and Sophie's Caves. Carnivores adapted to specialize in feeding on cave bears in response to the absence of large mammoth steppe game fauna such as mammoths, rhinos, bison, and horses. **B.** Mortality rates for cave bears, lions, and hyenas in the Zoolithen Cave as evidence of their use of the cave as a dwelling place or for other purposes only (from [3]).

bear den context within different landscapes (Fig. **14**) ranging from mammoth steppe lowlands (Westeregeln open air site [60, 61]) over middle high elevated mountains margins/valleys (Teuflskammer Cave [62], Perick Caves [20, 27, 63]) to the higher parts of the boreal forest mountains such as Upper Franconia [Zoolithen Cave [4, 5, 10, 38], Sophie's Cave [1 and herein]) clarifies the predator and guild relationships, which can be used even as a model for central Europe [3, 21]. Hyena den bone assemblages are very useful partly and indirectly for the "landscape reconstruction" by using the biodiversity and NISP of their prey species within the bone asssemblages [21, 34]. One of the best guild indicators are the mammoth/woolly rhinoceros/horse and the cave bears – which are very different represented within their bone amounts in caves, respectively *Mammuthus / Colodonta / Equus* only in hyena dens (Fig. **14A**) [21] (cf. Table **1**). The absence of any mammoth steppe megafauna within the Sophie's Cave from the MIS 5d-3 supports the position of the cave bear den in a boreal forest palaeoenvironment, whereas a hyena den seems to have been absent. This cave represents the typical type of "seasonal cave dwelling predation" by hyenas, lions and well proven also directly wolves (by their bones and coprolites, see above and [1]). In such caves, the carcass and bone damage is quite high, as described above (see small cave bear taphonomy). The cave bear hunt and scavenging special-ization in such middle high mountainous boreal forest regions of Europe, such as Upper Franconia, resulted from the rare seasonally in the valleys migrating mammoth steppe guild [3, 21].

The cave bear hunt and scavenging was quite dangerous in the caves, but somehow easy during cave bear winter hibernation times [1 - 5, 21]. The mortality rates and age classes of the predators and cave bears demonstrate the different use of caves and ethology (Fig. **14B**, and Chapter 6, Fig. **12**). The predator curves in larger cave bear dens, are always with high mortality rates in grown up predators (except in hyena and wolf cub raising den areas), but opposite in the cave bears (see Chapter 5, Fig. **14A**) [21]. Rapid and very short-term cave dwellers were lions due to hunting/killing purposes of cave bears and cubs even in deep cave parts, which are the main reason for cave bears to hibernate as deep as possible in caves to protect themselves [3, 21]. The same is proposed for the smaller leopard felids, whereas their rare material in European caves and unique record of several skeletons up to 2 km deep within a cave bear den in Bosnia Herzegovina [6] does not allow a more detailed picture of their ethology concerning their probable rare cave bear hunt. Hyenas used caves as three different den types, but always only in the entrance areas: a. Cub raising den, b. Communal den, c. Prey storage den and dwelled into the cave bear dens for cave bear scavenging and less for hunting [21]. Wolves also used cave entrances or small cavities for cub raising purposes only and dwelled into the cave bear dens for feeding [1].

REFERENCES

[1] Diedrich C. Ice Age geomorphological Ahorn Valley and Ailsbach River terrace evolution– and its importance for the cave use possibilities by cave bears, top predators (hyenas, wolves and lions) and humans (Late Magdalénians) in the Frankonia Karst – case studies in the Sophie's Cave near Kirchahorn, Bavaria. Quat Sci J 2013; 62(2): 162-74.

[2] Diedrich C. Cave bear killers, scavengers between the Scandinavian and Alpine Ice shields – the last hyenas and cave bears in antagonism – and the reason why cave bears hibernated deeply in caves. Stalactite 2009; 58(2): 53-63.

[3] Diedrich C. Cave bear killers and scavengers from the last Ice Age of central Europe: Feeding specializations in response to the absence of mammoth steppe fauna from mountainous regions. Quat Int 2011; 255: 59-78.

[4] Diedrich C. The largest European lion *Panthera leo spelaea* (Goldfuss) population from the Zoolithen Cave, Germany – specialized cave bear predators of Europe. Hist Biol 2011; 23(2-3): 271-311.

[5] Diedrich C. The Late Pleistocene spotted hyena *Crocuta crocuta spelaea* (Goldfuss 1823) population with its type specimens from the Zoolithen Cave at Gaillenreuth (Bavaria, South Germany) – a hyena cub raising den of specialized cave bear scavengers in boreal forest environments of Central Europe. Hist Biol 2011: 1-33.

[6] Diedrich C. Late Pleistocene leopards across Europe - most northern European population, highest elevated records in the Alps, complete skeletons in the Dinarids and comparison to the Ice Age cave art. Quat Sci Rev 2013; 76: 167-93.

[7] Diedrich C. Ichnological and ethological studies in one of Europe's famous bear den in the Urşilor Cave (Carpathians, Romania). Ichnos 2011; 18(1): 9-26.

[8] Diedrich C. Extinctions of late ice age cave bears as a result of climate/habitat change and large carnivore lion/hyena/wolf predation stress in europe. ISRN Zoology 2013; 1-25.

[9] Diedrich C. Evolution, Horste, Taphonomie und Prädatoren der Rübeländer Höhlenbären, Harz (Norddeutschland). Mitt Verb dt Höhlen- Karstf 2013; 59(1): 4-29.

[10] Diedrich C. Holotype skulls, stratigraphy, bone taphonomy and excavation history in the Zoolithen Cave and new theory about Esper's "great deluge". Quat Sci J 2014; 63(1): 78-98.

[11] Rothschild BM, Diedrich C. Comparison of pathologies in the extinct Pleistocene Eurasian steppe lion *Pantherea leo spelaea* (Goldfuss 1810) to those in the modern lion, *Panthera leo* – Results of fights with hyenas, bears and lions and other ecological stress. Int J Paleopath 2012; 2: 187-98.

[12] Diedrich C. The Late Pleistocene *Panthera leo spelaea* (Goldfuss 1810) skeletons from the sloup and srbsko caves in czech republic (central europe) and contribution to steppe lion cranial pathologies and postmortally damages as results of interspecies fights, hyena antagonism and cave bear attacks. Bull Geosci 2011; 86(4): 817-40.

[13] Bocherens H, Drucker DG, Bonjean D, *et al.* Isotopic evidence for dietary ecology of cave lion (*Panthera spelaea*) in North-Western Europe: Prey choice, competition and implications for extinction. Quat Int 2011; 245: 249-61.

[14] Argenti P, Mazza PPA. Mortality analyses of the Late Pleistocene bears from Grotta Lattaia, central Italy. J Archaeol Sci 2006; 33: 1552-8.

[15] Debeljak D. Fossil population structure and mortality of the cave bear from the Mikrica Cave (North Slovenia). Acta Carsol 2007; 36(3): 475-84.

[16] Grandal-D' Anglade A, Vidal Romani JR. A population study of the cave bear (*Ursus spelaeus* Ros.-Hein.) from Cova Eiros (Triacastela, Galicia, Spain). Geobios 1997; 30(5): 723-31.

[17] Quilès J, Petrea C, Moldovan OT, *et al.* Cave bears from P. cu Oase (Banat, Romania): taphonomy and paleobiology. Compt Rend Acad Sci Palevol 2006; 5(8): 927-34.

[18] Rabeder G, Nagel D, Pachem M. Der Höhlenbär. Thorbecke Species 4,Thorbecke Stuttgart 2000; p. 111.

[19] Pinto Llona AC, Andrew PJ. Scavenging behaviour patterns in cave bears *Ursus spelaeus*. Rev Paléobiol 2004; 23: 845-54.

[20] Diedrich C. Die oberpleistozäne Population von *Ursus spelaeus* Rosenmüller 1794 aus dem eiszeitlichen Fleckenhyänenhorst Perick-Höhlen von Hemer (Sauerland, NW Deutschland). Philippia 2006; 12(4): 275-346.

[21] Diedrich C. Palaeopopulations of Late Pleistocene top predators in Europe: Ice Age spotted hyenas and steppe lions in battle and competition about prey. Paleont J 2014; 1-34.

[22] Schaller G. The Serengeti Lion. A Study of Predator-Prey Relations. The University of Chicago Press, Chicago 1972; p. 494.

[23] White P, Diedrich C. Taphonomy story of a modern African elephant *Loxodonta africana* carcass on a lakeshore in Zambia (Africa). Quat Int 2012; 276/277: 287-96.

[24] Diedrich C. Late Pleistocene Eemian hyena and steppe lion feeding strategies on their largest prey—Palaeoloxodon antiquus Falconer and Cautley 1845 at the straight-tusked elephant graveyard and Neanderthal site Neumark-Nord Lake 1, Central Germany. Archaeol Anthropolo Sci 2014; 6(3): 271-91.

[25] Diedrich C. Mammoth scavengers in Europe - the Ice Age spotted hyenas and steppe lions and their feeding strategies on their largest prey. (in review).

[26] Kruuk H, The spotted hyena. A story of predation and social behavior. The University of Chicago Press, Chicago 1972; p. 352.

[27] Diedrich C. Cracking and nibbling marks as indicators for the Upper Pleistocene spotted hyaena as a scavenger of cave bear (*Ursus spelaeus* Rosenmüller 1794) carcasses in the Perick Caves den of Northwest Germany. Abh Naturhist Ges Nürnberg 2005; 45: 73-90.

[28] Diedrich C. "Neanderthal bone flutes" – simply products of Ice Age spotted hyena scavenging activities on cave bear cubs in European cave bear dens. Roy Soc Open Sci 2015; 2: 14002.

[29] Diedrich C. The Ice Age spotted *Crocuta crocuta spelaea* (Goldfuss 1823) population, their excrements and prey from the Late Pleistocene hyena den Sloup Cave in the Moravian Karst; Czech Republic. Hist Biol 2012; 24(2): 161-85.

[30] Diedrich C. Europe's first Upper Pleistocene *Crocuta crocuta spelaea* (Goldfuss 1823) skeleton from the Koněprusy Caves - a hyena cave prey depot site in the Bohemian Karst (Czech Republic) – Late Pleistocene woolly rhinoceros scavengers. Hist Biol 2012; 24(1): 63-89.

[31] Diedrich C, Žák K. Prey deposits and den sites of the Upper Pleistocene hyena *Crocuta crocuta spelaea* (Goldfuss 1823) in horizontal and vertical caves of the Bohemian Karst (Czech Republic). Bull Geosci 2006; 81(4): 237-76.

[32] Diedrich C. Periodical use of the Balve Cave (NW Germany) as a Late Pleistocene *Crocuta crocuta spelaea* (Goldfuss 1823) den: Hyena occupations and bone accumulations vs. Human Middle Palaeolithic activity. Quat Int 2011; 233: 171-84.

[33] Diedrich C. Late Pleistocene spotted hyena den sites and specialized rhinoceros scavengers in the Thuringian Mountain Zechstein karst (Central Germany). Quat Sci J 2015; 64(1): 29-45.

[34] Fosse P, Brugal JP, Guadelli JL, Michel P, Tournepiche JF. Les repaires d' hyenes des cavernes en Europe occidentale: presentation et comparisons de quelques assemblages osseux. In: Economie Prehistorique, Les comportements de substance au Paleolithique, XVIII Rencontres internationales d'Archeologie et d'Historie d'Antibes, Editions APDCA. Sophia Antipolis 1998; pp. 44-61.

[35] Musil R. Die Höhle "Sveduv stůl", ein typischer Höhlenhyänenhorst. Anthropos NS 1962; 5(13): 97-260.

[36] Ehrenberg K, Sickenberg O, Stifft-Gottlieb A. Die Fuchs- oder Teufelslucken bei Eggenburg, Niederdonau. 1 Teil. Abh Zool-Bot Ges 1938; 17(1): 1-130.

[37] Goldfuss GA. Osteologische Beiträge zur Kenntnis verschiedener Säugethiere der Vorwelt. VI. Ueber die Hölen-Hyäne (*Hyaena spelaea*). Nov Act Phys-Med Acad Caes Leopold-Carol Nat Curios 1823; 3(2): 456-90.

[38] Diedrich C. The largest European Late Pleistocene wolf population from the Zoolithen Cave (Bavaria, Germany), taxonomy, sexual dimorphism – and taphonomic contribution to the extinct European *Canis lupus spelaeus* (Goldfuss 1823) as cave bear scavengers. (in prep).

[39] Leonard JA, Carles Vilà C, Dobbs KF, Koch PL, Wayne RK, Van Valkenburgh B. Megafaunal extinctions and the disappearance of a specialized wolf ecomorph. Current Bio 2007; 17(13): 1146-50.

[40] Bibikow DI. Der Wolf - *Canis lupus*. Wittenberg, Neue Brehm-Bücherei 2003; p. 198.

[41] Ziemen E. Der Wolf, Verhalten, *Ökologie und Mythos. Stuttgart*, Kosmos-Verlag 2003; p. 447.

[42] Baja I, Miguel FJ, Bárcena F. The importance of crossroads in faecal marking behaviour of the wolves (*Canis lupus*). Naturwissensch 2004; 91(10): 489-92.

[43]　Diedrich C. Typology of Ice Age spotted hyena *Crocuta crocuta spelaea* (Goldfuss 1823) coprolite aggregate pellets from the European Late Pleistocene and their significance at dens and scavenging sites. New Mex Mus Nat Hist Sci Bull 2012; 57: 369-77.

[44]　Rogers LL, Mech D. Interactions of wolves and black bears in northeastern Minnesota. J Mammal 1981; 62(2): 434-6.

[45]　Güttinger R. *Mustela erminea*. Denkschr Schweiz Akad Naturwissensch 1995; 103: 372-82.

[46]　Heptner VG, Sludskii AA. Mammals of the Soviet Union. Vol. II, part 1b, Carnivores (Mustelidae and Procyonidae). Washington DC: Smithsonian Institution Libraries and National Science Foundation 2002; p. 2142.

[47]　Reynolds SH. A monograph of the British Pleistocene Mammalia. The Mustelidae. Monogr Palaeont Soc London 1912; 2(4): 1-28.

[48]　Galik A. Die Größenvariation der pleistozänen Mauswiesel (*Mustela nivalis* L.) und Hermeline (*Mustela erminea* L.) (Musteliden, Mammalia) aus der Schusterlucke im Kremstal (Waldviertel, Niederösterreich). Wissensch Mitt Niederösterr Landesmus Wien 1997; 10: 4-61.

[49]　Reichstein H. Schädelvariabilität europäischer Mauswiesel (*Mustela nivalis* L.) und Hermeline (*Mustela erminea* L.) in Beziehung zu Verbreitung und Geschlecht. Z Säugetierk 1975; 22: 151-82.

[50]　Reichstein H. Beitrag zur Kenntnis des Sexualdimorphismus von *Mustela nivalis* und *M. erminea* Linné, 1758 nach Untersuchungen an postcranialen Skeletten aus Schleswig-Holstein. Ann Naturhist Mus Wien 1986; 88/89: 293-304.

[51]　Eberlein C. Die Musteliden aus drei Höhlen des Frankenjura (Zoolithenhöhle, Geudensteinhöhle und Höhle bei Hartenreuth). Unpublished diploma-thesis, University Erlangen 1996.

[52]　King CM. The life history strategies of *Mustela nivalis* and *Mustel erminea*. Acta Zool Fennica 1983; 174: 183-4.

[53]　Nowak RM. Walker's mammals of the world. 6 Auflage. Baltimore, Johns Hopkins University Press 1999; p. 2015.

[54]　Groiss JT. Paläontologische Untersuchungen in der Zoolithenhöhle bei Burggeilenreuth. Ein vorläufiger Bericht. Erlanger Forsch B Naturwiss 1979; 5: 79-93.

[55]　Diedrich C. Early Weichselian *Hystrix (Atherurus) brachyura* Linnaeus 1758 remains and chewed bones from the porcupine and hyena den cave Fuchsluken at the Rote Berg near Saalfeld (Thuringia, Germany). Open Palaeont J 2008; 1: 33-41.

[56]　Ranke J. Das Zwergloch und Hasenloch bei Pottenstein in Oberfranken. Beitr Anthropol Urgesch Bayerns 1879; 2: 209-10.

[57]　Heller F. Zur Diluvialfauna des Fuchsloches bei Siegmansbrunn, Landkr. Pegnitz. Geol Bl NO-Bayern 1955; 5: 49-70.

[58]　Nehring A. Über diluviale *Hystrix*-Reste aus bayrisch Oberfranken. Sitzungsber Ges Naturfr Berlin 1892; 10.

[59]　Kempe S, Al-Malabeh A, Döppes D, Frehat M, Henschel H-V, Rosendahl W. Hyena caves in Jordan. Scie Ann School Geol Arist Univ Thessaloniki (AUTH) 2006; spec vol 98: 201-12.

[60]　Diedrich C. Late Pleistocene *Crocuta crocuta spelaea* (Goldfuss 1823) clans as prezewalski horse hunters and woolly rhinoceros scavengers at the open air commuting den and contemporary Neanderthal camp site Westeregeln (central Germany). J Archaeol Sci 2012; 39(6): 1749-67.

[61] Diedrich C. Impact of the German Harz Mountain Weichselian ice-shield and valley glacier development onto Palaeolithics and megafauna disappearance. Quat Sci Rev 2013; 82: 167-98.

[62] Diedrich C. The *Crocuta crocuta spelaea* (Goldfuss 1823) population and its prey from the Late Pleistocene Teufelskammer Cave hyena den besides the famous Palaeolithic Neanderthal Cave (NRW, NW Germany). Hist Biol 2011; 23(2-3): 237-70.

[63] Diedrich C. Eine oberpleistozäne Population von *Crocuta crocuta spelaea* (Goldfuss 1823) aus dem eiszeitlichen Fleckenhyänenhorst Perick-Höhlen von Hemer (Sauerland, NW Deutschland) und ihr Kannibalismus. Philippia 2005; 12(2): 93-115.

CHAPTER 6

THE FINAL LATE PLEISTOCENE CAVE BEAR AND SPORADIC CARNIVORE (HYENA AND WOLF) DEN

Abstract: About 32.000-26.000 BP the largest cave bears *Ursus ingressus* Rabeder *et al.* 2004 used the Sophie's Cave such as other larger cave bear dens of the Zoolithen Cave, Große Teufels Cave and Geisloch Cave and others in Upper Franconia. At this time the large portal of the today's entrance was opened. In this hall and branching areas, the cave was used for denning and birth. The Ailsbach River terrace changed first with an elevation increasing that caused periodical floods of the anterior valley oriented cave part only. Within the partly dry cave, seasonal floods cleft two more fluvial sequences, which are dominated in the first stage by sands and gravels. In the last stage, "gravel/frost brekzia/glauconite sand till series" of the latest Late Pleistocene and around the LGM (app. 32.000-16.000 BP) the floods finally transported sediment and the bones only into the Ahornloch branching halls and Passages. The large cave bears were also scavenged and predated by the three top predators (lions, hyenas and wolves) that specialized especially in boreal forests on cave bear feeding as a result of rare and disappearing valley migratory mammoth steppe game. Hyenas used the Sophie's Cave only shortly as den in the Ahornloch Hall area and imported typical for cave dens in Europe some woolly mammooth *Mammuthus primigenius* (Blumenbach 1799), woolly rhinoceros *Coelodonta antiquitatis* (Blumenbach 1799), and *Equus caballus przewalski* Poljakov 1888 horse prey remains into the cave entrance halls, which bones show typical hyena caused bite/chew damage. Already before the climatic change not later then 24.000 BP, before the Last Glacial Maximum glacier extensions in Europe (LGM, 19.000 BP), with unsolved questionable "glacial signs" (?valley glaciers) in Upper Franconia and within the Sophie's Cave, caused the extinction of the last cave bears, their top predators, and most of the boreal forest megafauna in Upper Franconia and central Europe

Keywords: Final Late Pleistocene, sedimentology, terrace gravel infill, Ailsbach Valley geomorphology, glacial signs, largest cave bear species, cave bear den, bone taphonomy, predators and scavengers.

FINAL MIDDLE LATE PLEISTOCENE (32.000–24.000 BP, MIS 3-2) – LARGEST CAVE BEARS

The sections within the Sophie's Cave (Figs. **1-2**) show first a lowering of the Pre-Ailsbach River terrace within the later Late Pleistocene (app. 32.000-24.000 BP), on a level of about 400-415 m a.s.l [1]. Again, river terrace gravels and sands were washed into the anterior parts of the cave (Ahornloch and Clausstein Halls, Sand Chamber, and Bear Catacombs), whereas the gravel components are different to those of the Middle Pleistocene terrace sequence [1]. This results from the surrounding geomorphology change due to erosion history, because the river already eroded much deeper into the softer Early (or Black) Jurassic claystone series [1] (see Chapter 2, Fig. **1**). In total, again a river terrace sequence can be well seen in the first, fine-grained, and later gravel dominated deposits, whereas two phases are subdivided [1] (Fig. **1** and Fig. **13**).

Those fluvial river terrace cold period sediments are only present in the cave as "relic sediments" within the valley and contain only the large European cave bear species of *Ursus ingressus* [2 - 14] (Figs. **2-7**), which dates those layers with the P4 and other molar tooth morphology of multiple coned forms [2 - 3] between 32.000-24.000 BP due to comparisons of other radiocarbon and P4 dated dated *U. ingressus* cave bear remains from other European caves [2 - 9].

The preserved section (= sand-gravel sequence 2 [1]) is below the "metal floor plate" in the center of the main chamber of the Ahornloch Hall (Fig. **1-2**, Chapter 3, Fig. **1**). Above yellowish-white dolomite ash sands, a first frost brekzia is present (Fig. **2**) which also contains megafauna remains found within the section of *U. ingressus* – several teeth and bones, *Panthera leo spelaea* – phalanx II and III (see Fig. **8**), *Rangifer tarandus*-phalanx I (see Fig. **11**). After this cold period, during high-floods in warmer times, primary only sands were washed into the anterior halls by the meandering Prae-Ailsbach River [1]. This also had consequences on the large cave bear species skeletons, which were decomposed completely within the cave by periodic running water, which taphonomic situation is different to the older small cave bear forms (see Chapter 5).

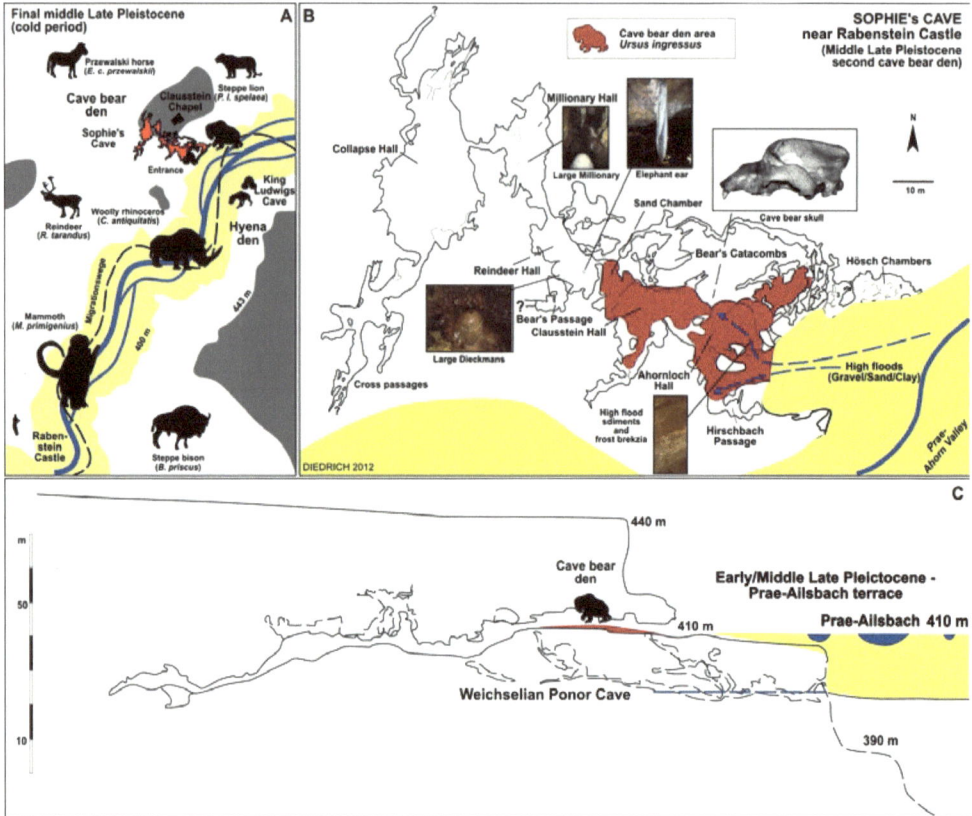

Figure 1. Final Middle Late Pleistocene (app. 32.000-24.000 BP, MIS 3-2) – large cave bears *Ursus ingressus* Rabeder *et al.* 2004 and Ailsbach River terrace elevation increase (modified from [1]).

The Ahornloch Hall, Clausstein Hall, and Sand Chamber of the Sophie's Cave were the den areas of the last and large cave bears of this region [1], such as the Zoolithen Cave [12], or by the author own unpublished observations, the Große Teufels Cave or Zahnloch Cave. Those largest European cave bears entered at that time what is the today's entrance, as a result of the lowered river terrace [1]. Although there must have been seasonal high floods (spring snow melting waters) the cave was still used for hibernation and as a cub raising site over some thousands of years between 32.000-24.000 BP [1], which explains larger amounts of sibling milk and not full developed cub permanent teeth (Fig. **6**) within the sands/gravels/brekzia, whereas hibernation nests were destroyed fluvially.

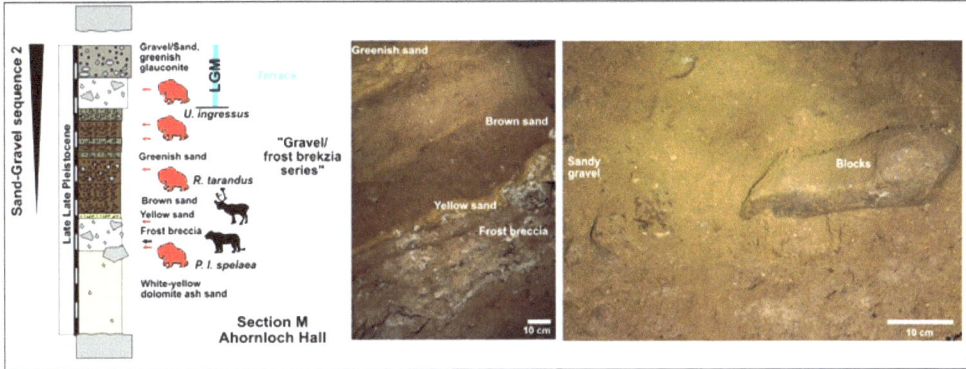

Figure 2. A. Section of the "Gravel/frost brekzia series" of the late Late Pleistocene and LGM (app. 32.000-16.000 BP, MIS 3-2) in the Ahornloch Hall. Lower part of the section with frost brekzia and above LGM sand-gravel sediments of greenish (glauconite) to brown sandy layers with larger not well rounded blocks (up to 30 cm) (modified from [1]).

The Largest Cave Bear Species *Ursus Ingressus*

Those large cave bears were found, in contrast, in the anterior cave areas (Sand Chamber, Bear's Catacombs, Clausstein and Ahornloch Halls) [1] and are even represented with few survived historical skull finds (Figs. **3-7**). The much larger last European cave bear species *U. ingressus* Rabeder *et al.* 2004 [2 - 14] is represented also by a single lower jaw (Fig. **6D**), that was found cemented into the youngest speleothem layer, along with the reindeer antlers. It dated in Europe also into the late Late Pleistocene between 32.000-24.000 BP, MIS 3-2 (Fig. **7**), according to more irregular enamel tooth surfaces and multiple coned P4 teeth (see Chapter 4, Fig. **14**), and the C^{14} data of other finds in Europe [2 - 14]. The occurrence of this along with the reindeer antlers dates both not older than 32.000 BP (see fitting reinder antler C^{14} date in Chapter 7 = 30.830-30.340 cal. BP), and additionally dates biostratigraphically the smaller cave bears *U. s. ersmus/spelaeus* of the bone field older than about 32.000 BP (see Chapter 3, Fig. **1**-Section F). This mandible of a male with typical sexual dimorphism large canine must have been deposited, when the cave was flooded in the final late Late Pleistocene (Fig. **13**), similar to how most bones of this species were transported in 99% fluvial only into the anterior cave parts, especially the Bear's Catacombs, but also lower chambers below the Ahornloch Hall [1].

Figure 3. The last and largest cave bear species *Ursus ingressus* with its typical stepped frontal and high shoulder also due to hind limb shortening (Illustration G. "Rinaldino" Teichmann).

The skulls (Fig. **4**, Fig. **6**) and postcranial bones (Fig. **5**), of the largest and final European cave bear species *U. ingressus* were found within the Sophie's Cave, but those have been transported or damaged by two factors. First, large predators (lions, hyenas, wolves) again scavenged their carcasses (see below carnivores: Fig. **8**) which explain also the lack of articulated skeletons, and again higher amounts of bite damaged and cracked cave bear bones. In a secondary taphonomic process the single bones were washed during quick flood events into deeper cave parts, mainly into the Bear Catacombs (Fig. **1**) [1]. The best preserved examples of water transported bones and sediments are found within the Bear Catacombs (Fig. **1**) [1]. There, only a few rounded and larger-sized (3-8 cm in size) dolomite gravels and few surface water-polished bones accumulated in non-stratified sediment gravels (no section present, yet). The uppermost bones are even attached to the youngest speleothem layer, such as a half skull, tibia or pelvic bone only of the large *U. ingressus*. In the Bear Catacombs cub remains were saved within the "modern bone dump" (made in 1975 by cavers, when also 6 skulls disappeared, see law conflict in Chapter 1). Sibbling, cub to subadult remains are present in abundance, especially with milk teeth which are found in the dolomite ash sands in the Ahornloch Hall (Fig. **6A**). Also skulls and skull fragments are present in the anterior cave areas from different aged sibblings and cubs or subadults ranging and from approx. 3-12 months in age (Fig. **6A-C**).

The postcranial bone material has several cub longbones mainly (Fig. **6B**), due to "anthropogenic selection" (bone dump, bone depot behind Sand Chamber). The few grown up individual cave bear remains (Fig. **5**) allow no further osteometric analyses concerning their sexual dimorphism. However, directly compared to *U. ingressus* bone material from the Hermann's Cave of the Harz Mountain Range [13], the few complete bones can be attributed either to larger proportioned males or smaller females, which is best to determine on the larger bones such as scapula, humerus, ulna, radius or pelvis, femur, tibia and fibula (Fig. **5**). Also pedal bones are different in sizes, which can be demons-trated on the Sophie's Cave material only with future excavated bones from the anterior cave areas.

Only a single pathologic thoracic vertebra (Fig. **5**) with exostoses does not allow a further discussion, but this pathology is typical and similar as described for the small cave bear subspecies (see Chapter 4).

In total, the bone amount of 453 bones (including two nearly complete skulls, only, Fig. **4** and Fig. **6**, Table. **1**) of *U. ingressus* of the Sophie's Cave are too few to "composite skeletons" for presentations within the show cave.

For the Sophie's Cave and the region most important are the different Late Pleistocene biostratigraphic occurrences of *U. ingressus* and the older *U. s. eremus/spelaeus* cave bears, which were not sympatric in the Sophie's Cave. In other southwestern German Swabian Alb caves, those are believed to have lived in sympatry [9], possibly only short-term after the immigration of the largest cave bears. The Sophie's Cave is the only cave in Upper Franconia with a good skull based, such as P4 and partly C^{14} dated biostratigraphy. The Zoolithen Cave [13], and Große Teufels Cave have all bones mixed from different species/subspecies and time periods, and are not sufficient for the evolutionary/migratory or ecological question in detail to solve (Fig. **7**), although those contain even more bones [13].

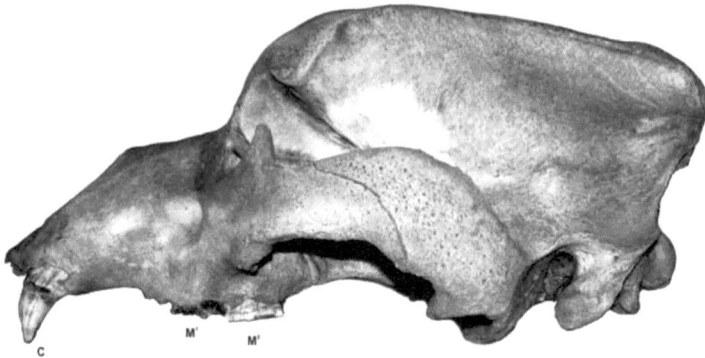

C   M'   M'

10 cm

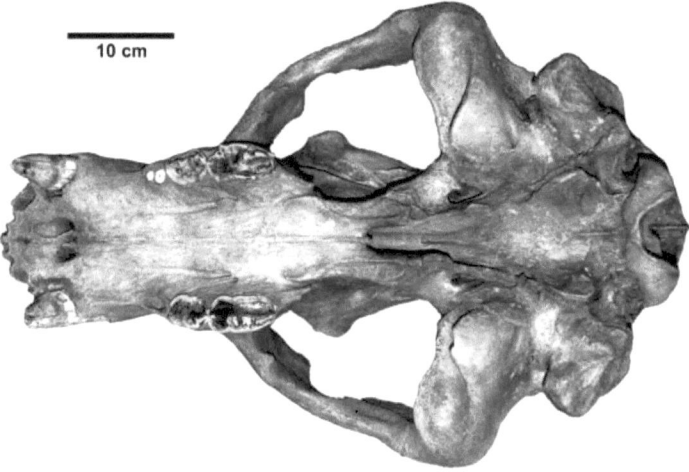

Cranium

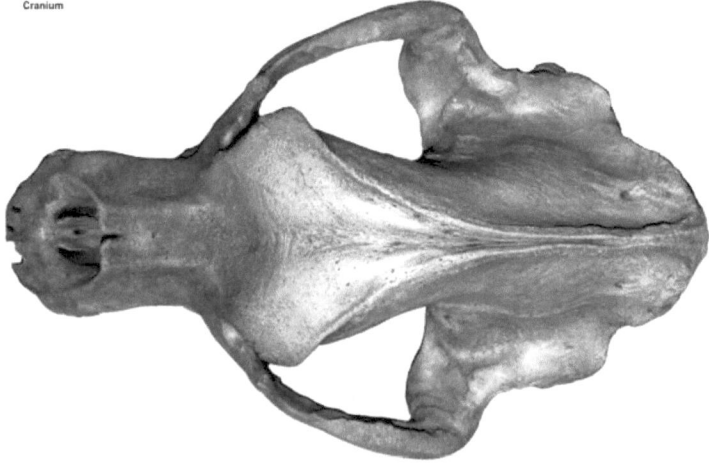

Figure 4. Skull of a 20-25 years old senile male cave bear *U. ingressus* from the "anterior cave part" of the Sophie's Cave with few lowering and high sagittal (as an adaptation of vegetarian nutrition) (coll. Urweltmuseum Oberfranken Bayreuth).

Whereas in the Sophie's Cave there is herein presented and confirmed a known evolutionary trend (smaller cave bears: *U. s. eremus* to *U. s. spelaeus* "forms/haplotypes/ subspecies") and possibly an extinction of those within a climatic change [2 - 16], the possible intrusion of the largest cave bears *U. ingresus* possibly from the East (Carpathians) proposed [12], still remains open to several questions.

Table 1. Late Middle/Late Late Pleistocene megafauna *U. ingressus* time bone assemblage (NISP = 482) from the anterior Ahornloch/Claustsein/Bear's Catacombs of the final Late Pleistocene *U. ingressus* bonebeds (app. 32-000-24.000 BP). This area was a sporadic/seasonal short-term use Ice Age spotted hyena *Crocuta crocuta spelaea* (Goldfuss 1823) and Ice Age wolf *Canis lupus spelaeus* (Goldfuss 1823) den, but mainly cave bear den of the last large cave bears.

Species	Bone amount = NISP (represented bone types)
Ursus ingressus	453 (2 skulls, jaws, teeth and postcranial bones)
Panthera leo spelaea	3 (Mandible, 2 phalange)
Crocuta crocuta spelaea	1 (Mandible fragment)
Canis lupus spelaeus	15 (Postcranial bones)
Vulpes lagopus	1 (Praemaxillary)
Mammuthus primigenius	2 (Coxae)
Coelodonta antiquitatis	2 (Humeri)
Equus caballus przewalskii	3 (Teeth)
Rangifer tarandus	2 (Phalangae)

As mentioned in Chapter 4 – *U. ingressus* might be simply the younger synonym of *Ursus spelaeus spelaeus* Rosenmüller 1794 [4 - 9, 16], if comparing by morphology the still DNA untested holotype [12, 17, 18]. Solving this main question first, will possibly not clarify the evolution, but will change the taxonomical nomenclature possibly ("*U. ingressus*" is then *U. s. spelaeus* and "*U. ingressus*" has to receive another name), such as the migration and origin of the largest European cave bears.

The extinction of the largest last cave bears of Europe around 24.000 BP is well established by C^{14} data [4 - 9, 16], also herein within the biostratigraphy and newest test of the P4 morphology and skull shape correlation to other sites such as the Hermann's Cave [13]. Its extinction is clearly a chain reaction and combination of: 1. climate change with coldest climate within the LGM at 19.000 BP [19], with partly newest discovered glaciations of even middle European mountain peaks such as the Harz Mountain [13, 20]. 2. Vegatation change due to the climate change within the middle high mountain boreal forests and less nutrition = high alpine vegetation and less blueberry-vegetation [12, 13, 20]. 3. Rising predation stress by top predators such as lions, leopards, hyenas and wolves due to extinction of the mammoth steppe megafauna [21 - 25]. 4. Predation stress with the new arriving Cromagnon humans about 35.000 BP (Aurignaciens) which started cave bear hunting in caves which continued at minimum into the Gravettian [26, 27] (see also Chapter 7, also possibly Gravettians hunted in the Sophie's Cave large cave bears).

Lions as Main Cave Bear Cub Killers

The only bone proof within the Sophie's Cave of the steppe lion *Panthera leo spelaea* (Goldfuss 1810) [24, 28, 29] was discovered new in 2010 in the Ahornloch Hall (two phalangae: Fig. **8**). The claw bone (phalanx III) was found in the surface sediments (yellow-white dolomite ash sands) of a side branch of the Ahornloch Hall, together with large cave bear *U. ingressus* teeth and bones (= in the cave bear bonebeds). Also a second phalanx bone, again of a subadult animal was found within the section together with *U. ingressus* bones in layer 4, the first and lowermost frost brekzia layer in the Ahornloch Hall (Fig. **2**). Furthermore, there is a historical find of a left mandible (Fig. **8**) kept in the University of Munich collection, which was misinterpreted, due to its small size (= juvenile, see also comparable juvenile jaws from the Zoolithen Cave in same sizes [24]) as belonging incorrectly to another felid "*Panthera onca*", which is indeed, an older Early-Middle Pleistocene large felid [30] not being proven in Upper Franconia yet. All three steppe lion remains belong to one or some adolescent lion(s) individual(s) (Fig. **8**).

Such a subadult carcass (of which only few material has been found) might even result from hyena import, which is known for other hyena dens (Perick Caves,

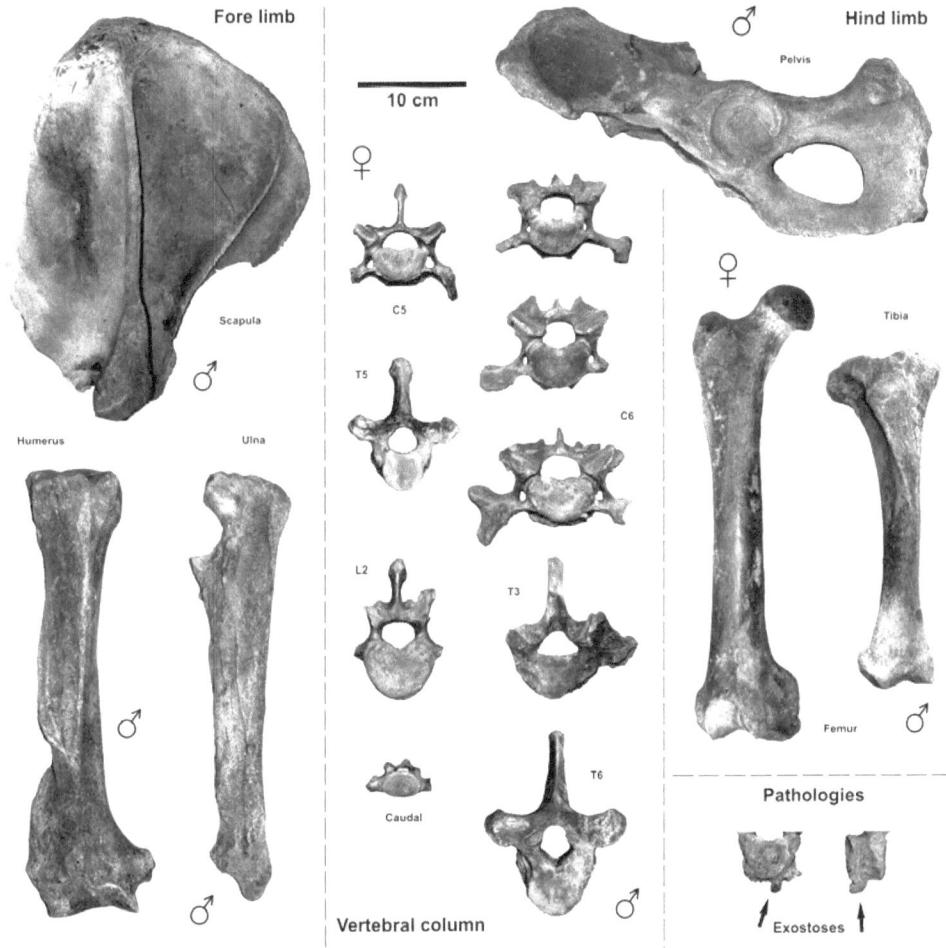

Figure 5. Postcranial bones of adult females and males of the large cave bear *Ursus ingressus* from the anterior cave areas, the Ahornloch Hall and mainly the Bear's Catacombs of the Sophie's Cave (coll. Rabenstein Castle Museum).

Germany) at cave bear den entrances or at open air den sites [31, 32]. Possibly a young lion wanted to steal hyena prey at their den site (Ahornloch Hall area), in which a weak subadult lion was finally killed during a conflict or battle in the cave entrance by hyenas, similiar as proposed for other European hyena dens, such as the Srbsko Chlum Cave, Czech Republic [33]. Both scenarios would support the presence of a periodic/seasonal/short-term hyena den use of the anterior cave areas (Ahornloch Hall area). A final scenario of the lion's death might have been a lost battle against an adult cave bear during cave bear cub predation (Fig. **12B**).

As known after studies in several European caves and at many open air sites, the steppe lions of the late Pleistocene were not cave, but mammoth steppe or boreal forest inhabitants [1, 12, 13, 15, 20, 21, 23, 24, 31 - 35], similar to modern lions in Africa who live solely in open savannah or steppe environments, and never in caves [36]. The Ice Age steppe lions were only periodic cave dwellers in middle high mountainous boreal forest areas and never even used cave entrances for cub protection, simply because those areas were also occupied by Ice Age spotted hyenas or Ice Age wolves [1, 12, 13, 20] (Fig. **12**), or were even used as camp or burial sites by Neanderthals/Cromagnon humans [37, 38]. In the Early/Middle Late Pleistocene middle high mountainous boreal forest regions, such as Upper Franconia, those lions could not kill abundant mammoth steppe megafauna prey biomass [15], and even less during seasonal migrations within the river valleys, such as the Ailsbch and Wiesent Valleys (Fig. **1**). This is best documented by the absence/rare mammoth steppe megafauna remains which few bones have been found at the largest Upper Franconia Ice Age spotted hyena den Zoolithen Cave [12], or the Sophie's Cave hyena den within the Ahornloch area.

The only abundant "meat colossus" of boreal forests were the cave bears, easy to kill during hibernation times, even if deeper in caves [21, 23, 24]. The hunt of cave bears by steppe lions was recently also proposed within the Zoolithen Cave of Upper Franconia [12, 24] (Fig. **12**). A hunt even deeper within cave bear den caves is indicated by a. The presence of lion remains (often articulated skeletons, if autochthonous non-flood situation of bonebeds [21, 23, 24]) in caves, and b. Large, but few bite impact damaged cave bear longbone joints. Lions (single or in packs, which remains unsolved also towards the sex composition) killed cave bears, especially cubs [25], in times when food was ample in the boreal forests [21, 23, 24]. Time by time, a lion must have been killed in a battle with an adult father or mother cave bear protecting their cubs (Fig. **12B**) [21, 23, 24].

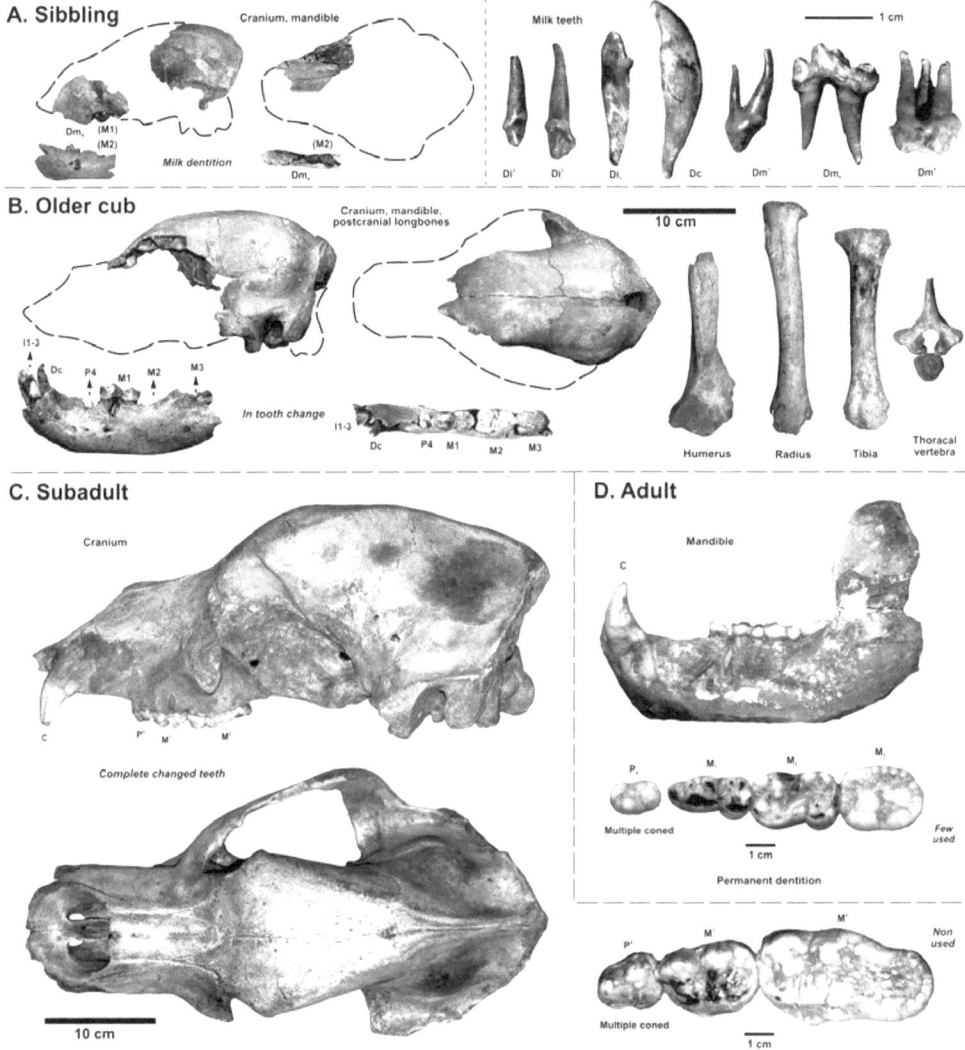

Figure 6. Skull and postcranial bone remains of different aged siblings to grown up *U. ingressus* cave bears from the anterior cave areas (Clausstein/Ahornloch Halls and Bear Catacombs, such as Reindeer Hall). **A.** Composed sibling skull, teeth and postcranial bone remains of several cave bear cubs, all from the Clausstein/Ahornloch Halls. **B.** Older cub in age of tooth change from milk to permanent dentition. Skull, mandible and postcranial bones from different individuals of the Clausstein/Ahornloch Halls. **C.** Subadult individual skull (in "spelaeoid cave bear cranial shape" with angled saggital crest). **D.** Left mandible of an adult male from the uppermost speleothem layer (above *U. spelaeus* subsp. layer). (all coll. Rabenstein Castle Museum, except C.: coll. Urweltmuseum Oberfranken Bayreuth).

A convincing argument for the seasonal cave dwelling and cave bear hunt is also found in the age structure of the steppe lions which skeletons/bones were found within cave bear dens [21, 23, 24]. Lion cub remains (0.1%) are missing in cave bear dens, and are present extremely rare only at hyena den cave site prey bone accumulations [21, 23, 24]. Very few (0.9%) older lion cub to subadult lion remains have been found in caves of Europe (Fig. **8**), and those bones partly showing bite damage are always in connection with hyena dens [21, 23, 24] similar found herein within the Sophie's Cave short-term hyena den (Fig. **8**).

In 99.9% of the lion remains, there is an average equal ratio of adult to senile lion/lioness bones, whereas this varies in each cave between a ratio of 1:3, 1:1 or 3:1 of males/females [21, 23, 24].

Ice Age steppe lions of Europe never used caves or their entrances for any denning or cub raising shelter purpose [21, 23, 24]. They possibly a. Sheltered short-term in the entrances which were competed by Ice Age spotted hyenas, Ice Age wolves and humans, b. Dwelled into the caves to steal hyena/wolf prey and c. Hunted seasonally in winter on hibernating cave bears [21, 23, 24]. This predatory pressure, especially on the last cave bears of Europe by large Ice Age predators mainly by well-climbing nocturnal steppe lions [21, 23, 24], but also other top predators, explains best, why cave bears hibernated as deep as possible in caves all over Europe, even in the most dangerous and deepest passages – to protect themselves [15, 21, 23], which has the most spectacular example being within the Romanian Urşilor Cave, where cave bears went over 1.800 m deep, climbing even within vertical shafts [23, 39, 40].

The Ice Age Spotted Hyena Den Ahornloch Hall

The Sophie's Cave (Ahornloch Hall area) is but about 40 m higher than the opposite König-Ludwig Cave's hyena den [41] (see Chapter 1, Fig. **1**), which must both have been accessible during the late Late Pleistocene (the Pre-Ailsbach River terrace was at that time deeper and the cave was mostly free and open, (Fig. **1** [1]). At both sites, the bones and possible "archaeological layers" were destroyed in modern historical times and shoveled *e.g.* in front of the König-Ludwigs Cave, of which Buckland complained, when he visited the cave after the

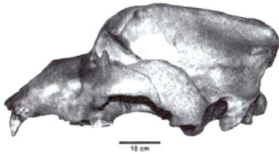

Ursus ingressus

Cave bear Evolution

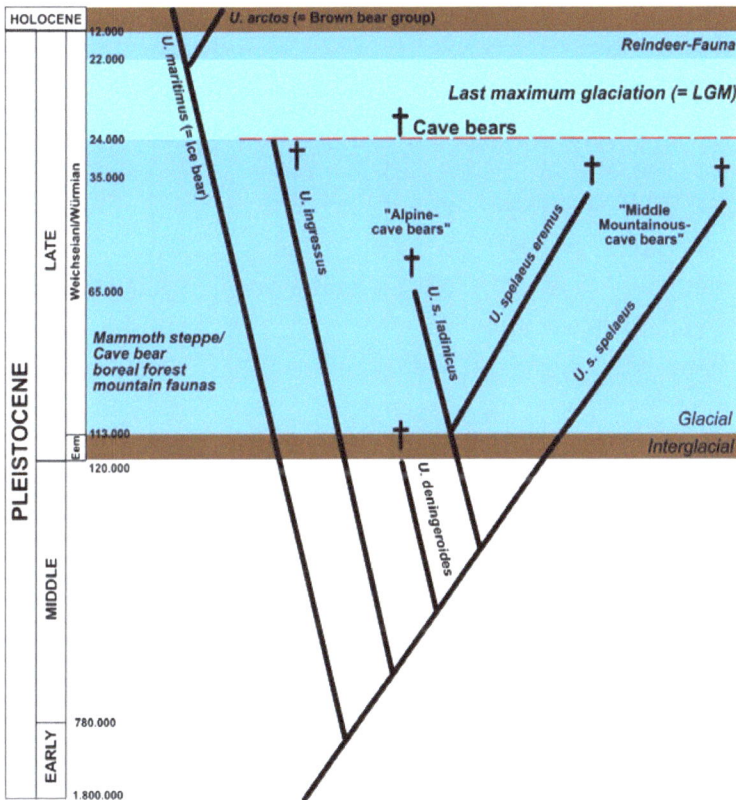

Ursus spelaeus eremus

Figure 7. Cave bear evolution and species/subspecies within the Late Pleistocene of Central Europe. While in the Bear's Passage, Reindeer and Millionary Halls small forms of *U. spelaeus* cf. *eremus* occurred in the early/middle Late Pleistocene, they were replaced by the larger *U. ingressus* forms which were found mainly in the Ahornloch or Clausstein Halls, Sand Chamber and Bear Catacombs (modified from [15]).

"destruction" of a hyena cave den site [41]. New bone finds of hyenas are missing in both caves from modern research times. Instead, undescribed cave bear remains of *U. ingressus* teeth were found recently in the König Ludwig's Cave indicating also the overlap with a latest Late Pleistocene cave bear den.

Hyena skull and bone remains must be expected in future explorations in the anterior Sophie's Cave parts, because the anterior part was a den including imported hyena prey remains of mammoth steppe fauna migratory prey. From this anterior part of the Sophie's Cave, a hyena skull was mentioned in the Graf zu Münster report [51], which was exposed at the Rabenstein Castle, today. This and other skulls are not to relocate anymore. At such hyena dens, the hyena bones generally count 10-35% of the NISP (excluding cave bear material, if present) in nearly all European hyena dens [42 - 52]. A mandible of the Ice Age spotted hyena *Crocuta crocuta spelaea* (Goldfuss 1823) [22, 42] (Fig. **8**) with an unclear origin from the "Kuhloch or Rabenstein Cave" being the Ahornloch Hall of the Sophie's Cave (or within the ?König-Ludwigs Cave, see Chapter 1) is housed today after being relocated by the author visits in the British Museum of Natural History in London. Perhaps this jaw is a relic of Buckland's research within both caves (König Ludwig's Cave and Sophie's Cave = Rabenstein Cave in his reports [41]), as he focused as a pioneer of the "modern hyena bone assemblage research ideas in caves" of Europe [41].

The hyena prey bone assemblage at the Sophie's Cave hyena den is limited and has only few bones of few mammoth steppe megafauna game (Figs. **9-10**). From the historical bone collections which came from the anterior halls of the Sophie's Cave, typically reddish colored bones are presented, which must have come from the Ahornloch Hall, and from there, the *U. ingressus* layers (32.000-24.000 BP, similar bone preservation and colour).

A mammoth coxa that was already mentioned historically from the upper part of the Reindeer Hall [41] seems to have been removed from the speleothem layer embedded somehow. A second coxa (Figs. **9-10**, possibly from same pelvis) was forgotten, because it was covered completely by the young speleothem layer. It was visible only by its acetabulum, which was beside the steps (Fig. **9**), and slightly damaged by a pickaxe during the installation of the visitor trail. The relatively small coxa of a young to mature adult animal has an acetabulum diameter of only 14 cm, which suites more female mammoths when compared to larger proportions of male bull skeleton pelvis or coxae acetabulae from Germany [53, 54].

After first rediscovery in 2010, and first cleaning on-site, old gluing of fractures with concrete was visible, such as from strong pickaxe damage. Due to conservatory reasons, the mammoth pelvic remain was completely removed. It was prepared, conserved with resin for future exposure. It was before sampled and C^{14}-dated 29.340-28.600 cal. BP (in the laboratory of Beta Analytic Inc., USA, no. SOPHIEMAM002, see also discussion Chapter 7), which date into the "*U. ingressus* or Gravettian" time of the late Middle Late Pleistocene or beginning of the MIS 2 within a short interstadial (= Greenland Interstadial 5 after [55], cf. Chapter 5, Fig. **14**), before a longer cold period started to increase to the Last Glacial Maximum (= LGM, sensu [19]). The pelvic remain from the Reindeer Hall has well visible bite damage pits from large carnivores which were left on typical places only large predators could produce. Two approximate nine millimeter round-oval wide, and several millimeters deep bite impact marks and elongated sinous bite scratches with similar large widths (4-8 mm) in the bone spongiosa of the coxa (Fig. **10**) might have been left either by canine teeth of *Crocuta* or their bone crushing premolars, or *Panthera* canine teeth (see Chapter 5, Fig. **4**). Canine impact marks of lions typically are found on those pelvic areas because they begin eating into the elephant carcass corpses from the anus (intestine/inner organ feeding [34]). Similar damages on the mammoth coxae are recently figured for several specimens of German open air or cave sites [56, 57].

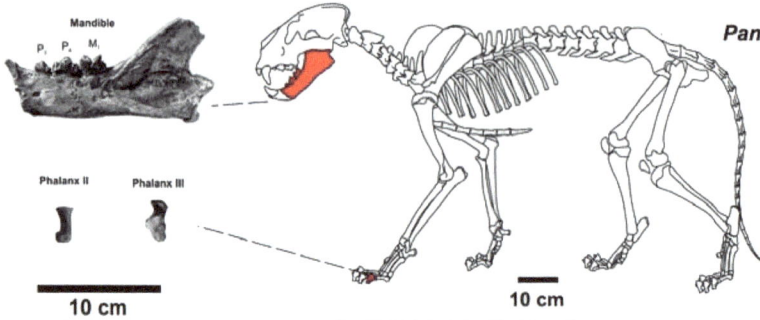

Steppe lion
Panthera leo spelaea
(Adolescent)

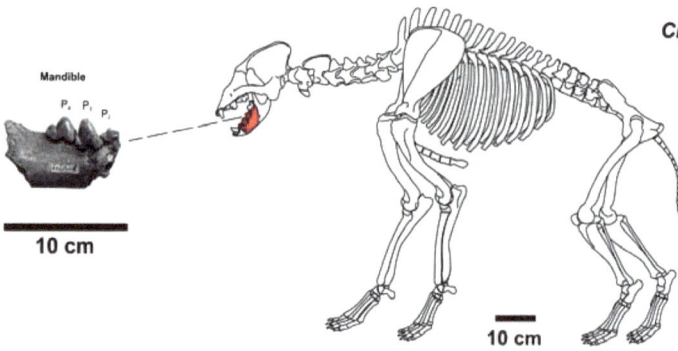

Mandible

P₃ P₄ M₁

Phalanx II   Phalanx III

10 cm   10 cm

Ice Age spotted hyena
Crocuta crocuta spelaea
(Adult)

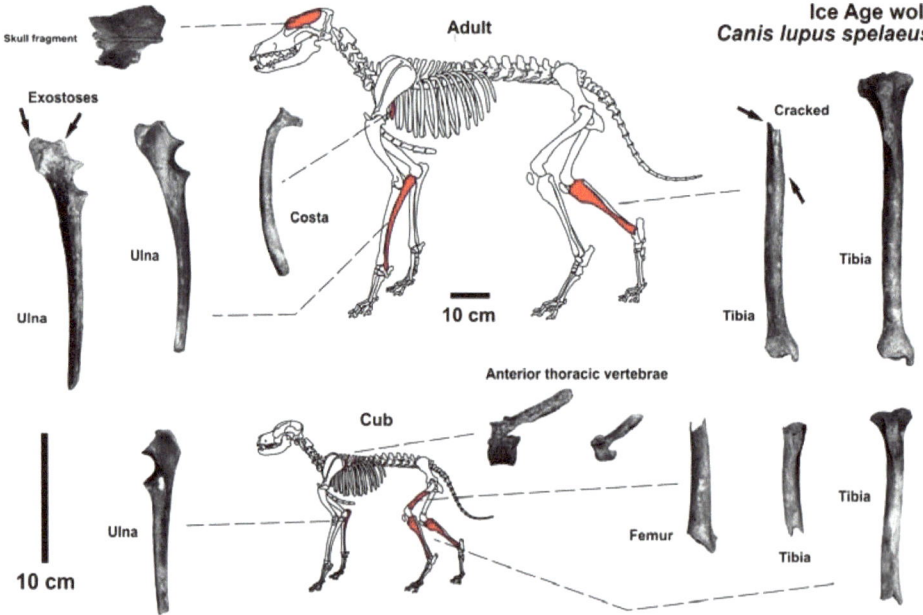

Mandible

P₂ P₃ P₄

10 cm   10 cm

Ice Age wolf
Canis lupus spelaeus

Skull fragment   Adult   Cracked

Exostoses

Ulna   Costa   Tibia

Ulna

Ulna   10 cm   Tibia   Tibia

Anterior thoracic vertebrae

Cub

Ulna   Femur   Tibia

Tibia

10 cm

Figure 8. Large cave bear *U. ingressus* Rabeder *et al*. 2004 top predator and scavenger remains, the steppe lion *Panthera leo spelaea* (Goldfuss 1810) bones of an adolescent animal (Ahornloch Hall area), phalanx III (Ahornloch Hall, section layer 4, see Fig. **1**) (coll. Rabenstein Castle Museum), and right mandible (Ahornloch Hall) (coll. BSPG no. 1894 I 501). Ice Age spotted hyena *Crocuta crocuta spelaea* (Goldfuss 1823) lower jaw (?from König Ludwigs Cave or most probably from the Ahornloch Hall of the Sophie's Cave, coll. British Museum Natural History London). Ice Age wolf *Canis lupus spelaeus* Goldfuss 1823 from the Bear Catacombs, Sand Hall, Reindeer Hall, Sophie's Cave (coll. Rabenstein Castle Museum).

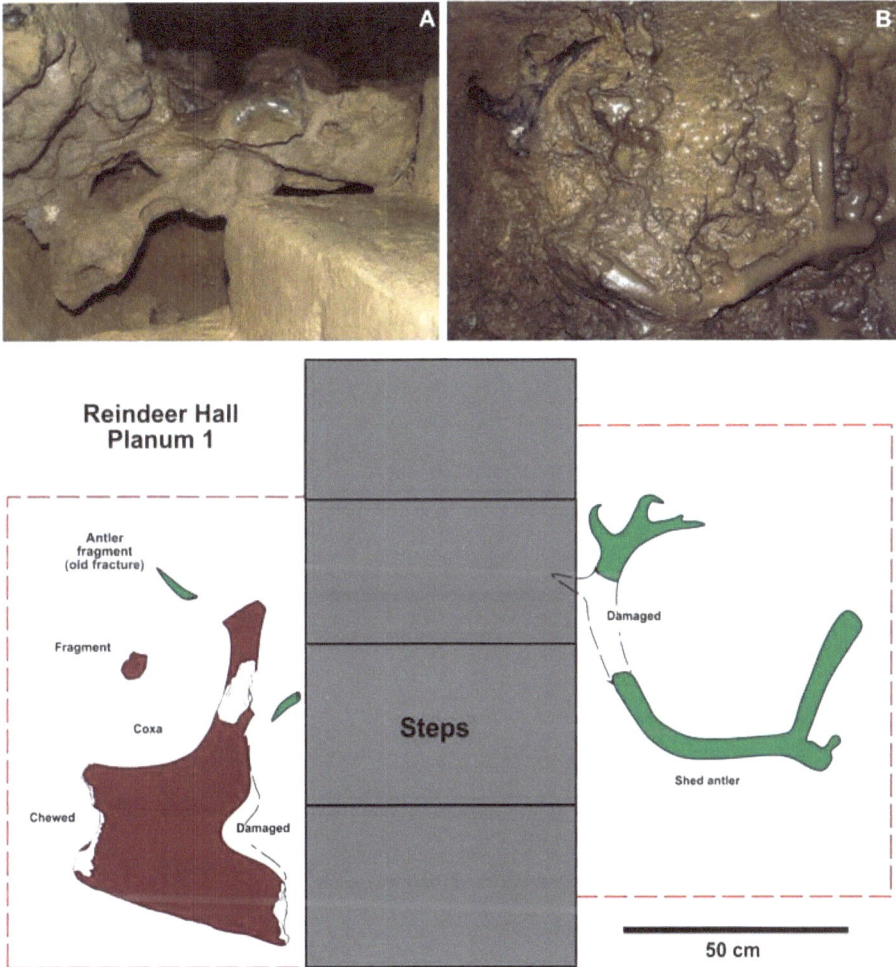

Figure 9. A. Mammoth pelvis coxa (C[14]-dated 29.340-28.600 cal. BP), and **B.** Reindeer shed antler (C[14]-dated 30.830-30.340 cal. BP, see also Chapter 7), such as distal reindeer antler branch fragments (with old break damages) in the middle part of the Reindeer Hall, besides the steps. Both have been damaged during the trail installation (coll. Rabenstein Castle Museum, antler *in situ*).

This is similarly reported from the Late Pleistocene straight tusk elephants of Neumark-Nord Lake in Germany [34], or on African elephants [58]. A lion fits easily through the pelvic canal, that is why they start feeding from there, and then the body cavity [58].

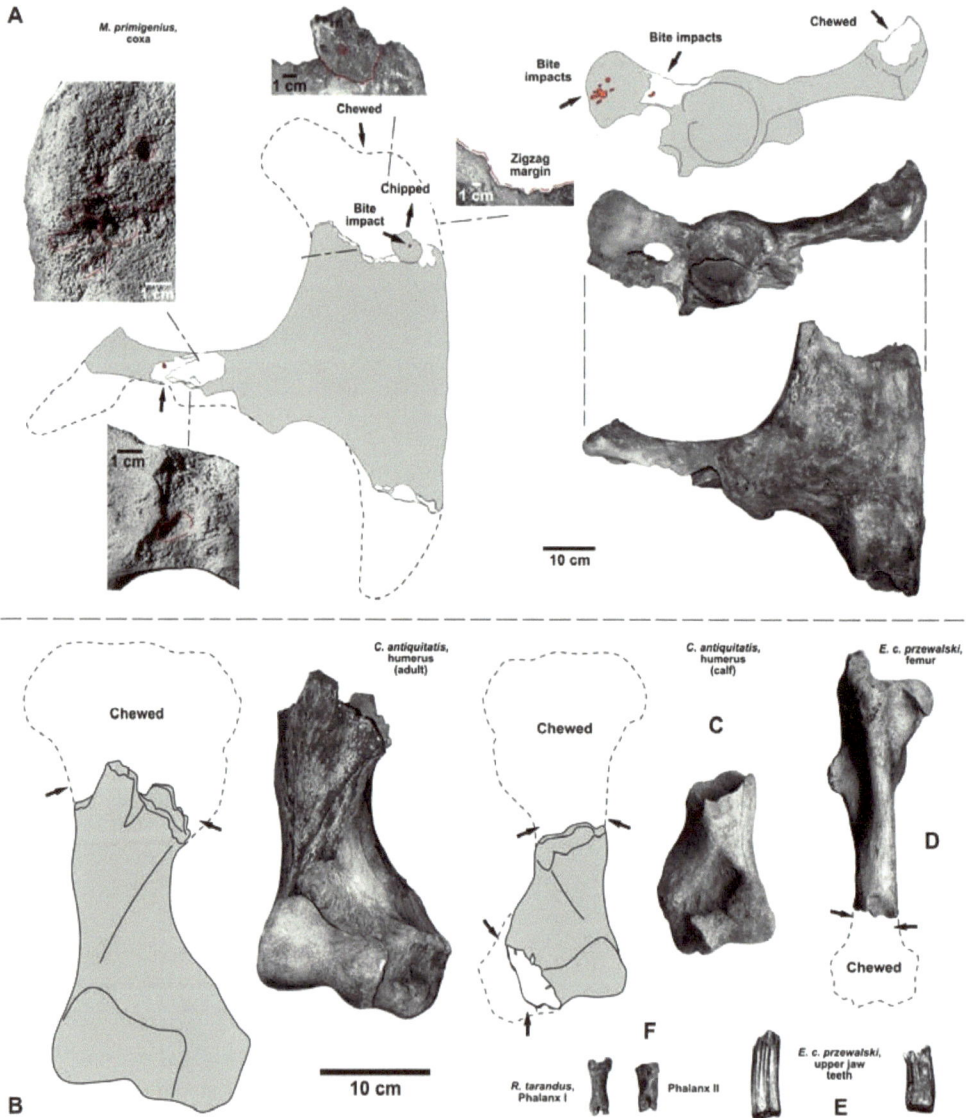

Figure 10. 1-2. Ice Age spotted hyena prey remains (mammoth, woolly rhinoceros, Przewalski horse, and reindeer) from the anterior cave area (upper Reindeer Hall, and mainly Ahornloch Hall and Clausstein Hall) which were imported and chewed. **A.** *Mammuthus primigenius* (Blumenbach 1799) coxa from the upper part of the Reindeer Hall (see position also Fig. **10**) with large carnivore (lion or hyena) bite impact marks of canines or hyena premolar crushing teeth (coll. Rabenstein Castle Museum). **B.** *Coelodonta antiquitatis* (Blumenbach 1799) grown up animal humerus with proximally by hyenas chewed joint (= damage stage 1) from the Ahornloch Hall area (coll. Urweltmuseum Oberfranken Bayreuth). **C.** *Coelodonta antiquitatis* (Blumenbach 1799) young animal humerus with proximally and distally by hyenas chewed joints (= damage stage 2) from the Ahornloch Hall area (coll. Urweltmuseum Oberfranken Bayreuth). **D.** *Equus ferus przewalskii* (Poljakoff 1888) distally by large carnivores chewed femur from the Ahornloch Hall area (coll. Urweltmuseum Oberfranken Bayreuth). **E.** *Equus ferus przewalskii* (Poljakoff 1888) upper jaw teeth from the Ahornloch Hall area (coll. Urweltmuseum Oberfranken Bayreuth). **F.** *Rangifer tarandus* Linnaeus 1758 phalanx I from the section (Fig. **2**) and phalanx II from the Ahornloch Hall area (coll. Urweltmuseum Oberfranken Bayreuth). Both must have passed through the stomach of a large carnivore (surface dissolution structures).

During scavenging, they must have damaged the soft parts on the distal areas of the pubis (Figs. **9-10**). Very important for the carnivore activity in the cave is a small, old fractured pelvic chip, which was excavated beside the coxa (Fig. **9**). This piece can be refitted to the chewed margin area of the pelvis, although this fragment has, in the center, a 61 mm wide bite hole (Fig. **10**).

The round-oval bite impact mark fits best to the premolar cracking teeth of hyenas (see Chapter 5, Fig. **4**) [59]. Those seem to have used, as modern hyenas do [60, 61], the premolars and last molar to cut and break bone pieces from the pelvic margin. It seems the mammoth pelvis was damaged possibly by both lions and hyenas, but first outside the cave, where they scavenged on the mammoth carcass in the river valley. The coxa must have been imported by hyenas then (if not possibly even by Late Palaeolithic Gravettian humans together with the reindeer antlers, see Chapter 7). Such mammoth pelvis pieces and other, larger mammoth bone remains were typically imported to several hyena cave dens in Europe, but there, pelvic remains are found finally as smaller and much more damaged pieces [47, 50, 51, 56, 57]. This coxa from the Sophie's cave has more the character of an open air site (den or carcass scavenging site) coxa find with typical initial bite damages, compared to open air/cave mammoth coxae [56]. Possibly, during the transport processes from the valley to the cave, hyenas left several bite impact marks behind the acetabulum area, where they were able to grab the coxa on its thinnest area (Fig. **10**). In the cave (however imported by hyenas or Gravettians), the one coxa was obviously further chewe damaged, thereby leaving a bone fragment nearby (Fig. **10**). Ice Age spotted hyenas used the anterior, or easy accessible cave parts, in the Sophie's Cave in maximum to the upper Reindeer Hall, for "prey food storage" at communal or prey storage dens against other hyena clans and lions [62].

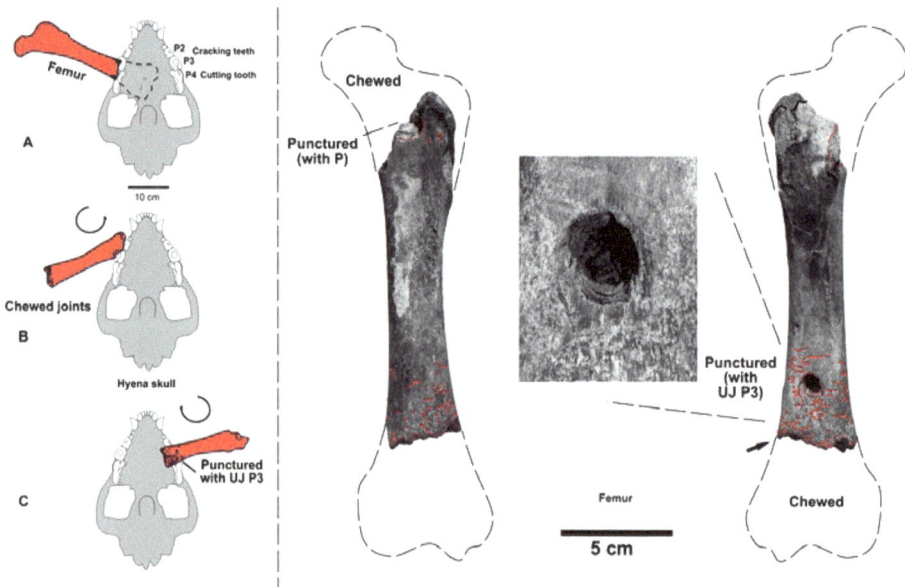

Figure 11. "Pseudo-Neanderthal bone flute", which is indeed a cave bear cub (*U. ingressus*) femur shaft from the Bear Catacombs. It was not cracked by hyenas, because there is no full calcification of the bone compacta (coll. Rabenstein Castle Museum).

Woolly rhinoceros *Coelodonta antiquitatis* (Blumenbach 1799) bones were also historically mentioned [41] and were rediscovered with two bones at least from the Sophie's Cave (Fig. **10**). Two very typical hyena chewed woolly rhinoceros humeri bones (Fig. **10**) are similarly found without the proximal joint, which is typical at many European Ice Age spotted hyena dens [44, 48, 51, 52, 63] and defined as "damage stages 1 and 2" for hyena damaged woolly rhinoceros humeri [59].

Horses were also mentioned with teeth [41], of which two were relocated in Bayreuth. There, also a distally chewed femur of a small cold period Przewalski's horse *Equus ferus przewalskii* (Poljakoff 1888) was found in the collection which have similar bone preservation as the woolly rhinoceros bones (Fig. **10**). Similar by hyenas mainly (partly also wolf) damagcd femorae were figured from other Late Pleistocene hyena dens of Europe [49, 64, 65].

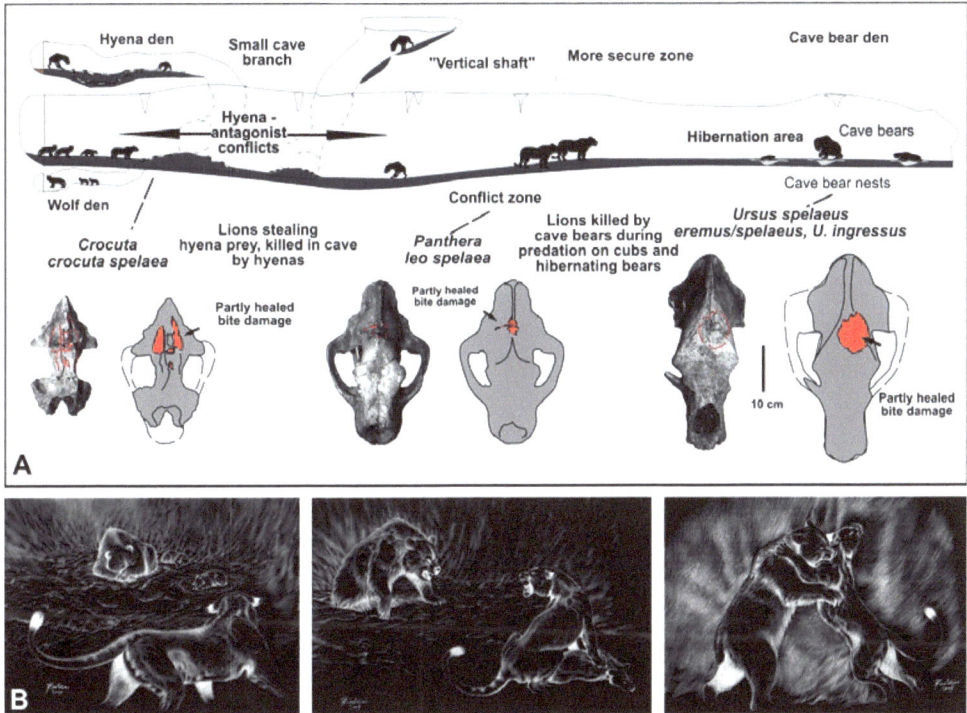

Figure 12. A. General Late Pleistocene model for the maximum of cave use and conflicts between all top predators and cave bears with pathologic bite trauma skull examples of the Zoolithen Cave, Upper Frankonia. This Late Pleistocene cave bear den model is similar for the Sophie's Cave. **B.** Steppe lions hunting cave bears during winter hibernation times in boreal forest mountain regions, here in the Sophie's Cave deepest hibernation area, the Millionary Hall ("cave imaging" illustrations by G. "Rinaldino" Teichmann, from [1]).

Reinder *Rangifer tarandus* Linnaeus 1758 remains are also rare in the Sophie's Cave material of the anterior cave area, which is different with the shed antlers of the Reindeer Hall (which are not of carnivore, but human depot origin: see Chapter 7). The bones are only distal leg elements (phalangae) typically found in much lesser amounts at Late Pleistocene Ice Age spotted hyena dens [62], whereas those also can be in context of a "wolf den" [66], which are not well studied for German caves (except herein see Chapter 5). The two reindeer phalanx bones from the section of the Ahornloch Hall (Fig. **10**) went through a "large carnivore stomach"is proven by their strong stomach acid surface corrosions.

The hyena none-cave bear game bone assemblage – which has obviously an exploring history impact - fit to the typical Late Pleistocene European Ice Age spotted hyena den prey bone assemblage, and taphonomy record [62], which is generally dominated by woolly rhinoceros *C. antiquitatis* [49, 51, 52, 59, 64, 65] and horse *E. f. przewalskii* remains, especially their teeth [49, 64, 65].

Those bone assemblages typically are found only in entrance areas of hyena den caves of Central Europe, and never deeper in caves [62]. The Sophie's Cave was situated directly on the river valley migration corridors of the mammoth steppe fauna and seem to have been used only for short-term or seasonally by hyenas as shelter (= communal den type [62]), but not for cub raising, nor as a larger game prey storage site. Similar as in other cave bear dens with hyena den overlap in the entrance areas in the surrounding caves such as Große Teufels Cave (unpublish own observations) and Zoolithen Cave [12], also in those, there are few remains (less then 1% of the non-cave bear bone NISP) of "mammoth steppe prey". The few material from the *U. ingressus* layers do not alow a well-based osteometric statistics such as presented for the smaller and older cave bear palaeopopulation (cf. Chapter 5, Fig. **14**). In general, also the *U. ingressus* bone NISP remains still similar, with dominance of cave bears (94%) and few other megafauna remains (6%) (see Table. **1**), reflecting still a continuing typical boreal forest cave bear dominated megafauna [62].

Ice Age spotted hyenas had no chance, similar as in many caves in Europe, to "move into" a cave bear den (at that time large *U. ingressus*, cf. carnivore/cave bear cave use model Fig. **12**) [21 - 23, 62]. At least over a few seasons or years, Ice Age spotted hyenas used the Ahornloch Hall for denning as mentioned, but those mainly dwelled therein to scavenge cave bear carcasses similar as known for many European caves [12, 13, 15, 21 - 23, 62, 68, 69], which is documented in the Sophie's Cave on several top predator bite damaged *U. ingressus* bones, even from material of the short-distance fluvial transported bones within the anterior cave and into the Bear Catacombs. The cave bear bones of *U. ingressus* with massive damage, especially on cub remains, support the dweller theory of the Anhornloch Hall by lions, hyenas and wolves as cave bear predators and scavengers [21 - 23, 62]. Most probably, hyenas cleaned the caves, mainly in spring-time, when old, young and weak bears had still not woken from hibernation [21 - 23, 62, 68]. Very typical indicators for hyena cave bear scavenging activities are the "pseudo-Neanderthal bone flutes" [68] (Fig. **11**).

Still today, some archaeologists believe those "carnivore produced damaged bones" are the "oldest musical instruments" [69], but actually, in most of the larger cave bear dens of Europe, similar damaged cave bear cub femorae have been found and are now recognized as always being in hyena context [68]. The damage history of especially cave bear cub femorae was explained in detail on the damage stages and with the comparison of the hyena dentition (Figs. 46, 49) [68]. Those first chewed the soft distal joints including the cartilage, using hereby the posterior "cutting teeth" (Fig. **11**). Further, they rotated the bone and cut the other side. Instinctively, hyenas tried then as with all other medium to larger sized prey long bones to crack the bones to reach the marrow, swallowing bone fragments [62]. The crushing of the shaft mostly was successful on bones from adult bears, which were fully calcified leaving hereby large amounts of longbone fragments at cave bear dens [68]. This cracking was unsuccessful on cub longbones (here femorae), because they were still soft and not fully calcified [68]. Hyenas were biting with their cracking molar teeth (P) on the "soft cub femur shafts" which did not break, leaving one (here Sophie's Cave example, Fig. **11**) or several round-oval bite impact impressions in the outer third, or even the middle of the bone shafts [67, 68]. Sometimes punctures were left on both sides, giving the impression of a "flute design" [68]. Besides those larger holes, and at the distal shaft ends, typically there are several triangular or elongated bite marks which resulted from the lower jaw P_4 and upper jaw M^1 teeth [68], as well documented herein at the Sophie's Cave cub femur (Fig. **11**). At cave bear den sites where those pseudo bone-flutes and other damaged bones are common, often hyena remains are missing (or are rare), because they were unable to occupy those caves in the entrances as their den (Fig. **12**) [21 - 23, 62, 67, 68].

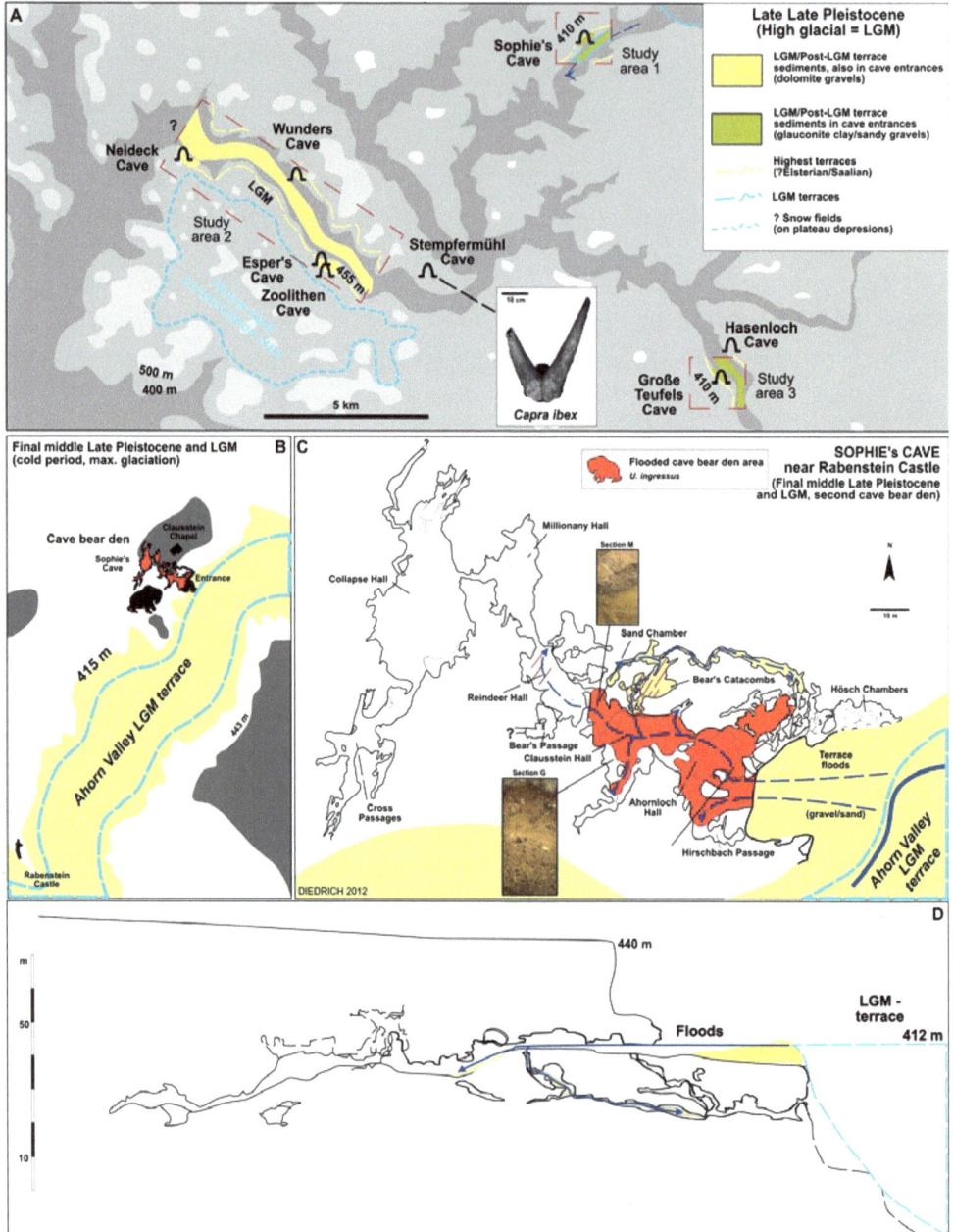

A

Sophie's Cave
410 m
Study area 1

Late Late Pleistocene
(High glacial = LGM)

LGM/Post-LGM terrace sediments, also in cave entrances (dolomite gravels)

LGM/Post-LGM terrace sediments in cave entrances (glauconite clay/sandy gravels)

Highest terraces (?Elsterian/Saalian)

LGM terraces

? Snow fields (on plateau depresions)

Wunders Cave

Neideck Cave
?

LGM

Study area 2

Esper's Cave
Zoolithen Cave
455 m

Stempfermühl Cave

10 cm

Capra ibex

Hasenloch Cave

Große Teufels Cave
410 m
Study area 3

500 m
400 m

5 km

B Final middle Late Pleistocene and LGM
(cold period, max. glaciation)

Cave bear den

Claustein Chapel

Sophie's Cave
Entrance

415 m
443 m

Ahorn Valley LGM terrace

Rabenstein Castle

C

Flooded cave bear den area
U. ingressus

SOPHIE's CAVE
near Rabenstein Castle
(Final middle Late Pleistocene and LGM, second cave bear den)

Millionany Hall

Section M

Collapse Hall

Sand Chamber

N

10 m

Bear's Catacombs

Hösch Chambers

Reindeer Hall

?
Bear's Passage
Claustein Hall

Section G

Cross Passages

Ahornloch Hall

Hirschbach Passage

Terrace floods

(gravel/sand)

Ahorn Valley LGM terrace

DIEDRICH 2012

D

440 m

m

50

Floods

LGM - terrace
412 m

10

Figure 13. A. Final middle Late Pleistocene and LGM (app. 24.000-LGM-16.000 BP) "glaciation" model for Upper Franconia based on cave relic sediments and megafauna of three main caves (Zoolithen Cave, Große Teufels Cave and Sophie's Cave). **B-D.** LGM valley terrace infill with fluvial transport of the older *U. ingressus* cave bear bonebeds on secondary position into deeper cave parts (modified from [1]). Alpine mountain goat *Capra ibex* Linneaus 1758 skull from around the LGM (app. 24.000-LGM-16.000 BP, MIS 2-1) which was collected in the small Stempfermühlen Cave (= LGM/Post-LGM brown bear den) not far from the Sophie's Cave (coll. Museum Tüchersfeld near Pottenstein).

A New Sporadic Wolf Den

As mentioned in a historical publication [41], a wolf skull was kept in the Rabenstein Castle in former times, being lost until today, together with many cave bear skulls. New wolf bone finds in the anterior cave area from the Bear Catacombs (from bone dump) and the diagonal shaft of the Ahornloch Hall (also reworked and on secondary position), such as the Sand Chamber (there *in situ* find of a vertebra within the final Late Pleistocene layer), and their relatively high amount on the bone NISP of the *U. ingressus* time (3%, see Table. **1**) indicate the presence of wolves, even cubs before the LGM (cf. Fig. **7**, Chapter 5, Fig. **14**). The Late Pleistocene European *Canis lupus spelaeus* Goldfuss 1823 [42, 70] wolves, which were started to be revised with the Zoolithen Cave paratype skull (holotype original is lost) [12] were compared to the modern arctic Canadian Timber wolf subspecies. Those extinct Pleistocene wolves are few smaller as Timber wolves, but are few larger as the modern European *Canis lupus lupus* Linnaeus 1758 [71]. The large bone proportions of the Late Pleistocene *C. l. spelaeus*, herein including for osteometric comparisons the only complete tibia of the Sophie's Cave material (Fig. **8**), are only about 1/8 larger than modern European *C. l. lupus* [71]. Those Late Pleistocene extinct glacial wolves are also present with bones in the Zoolithen Cave (more than 450 bones and several skulls) [12, 71], Großen Teufels Cave (few unpublished postcranial bones). This medium-large "ecomorph" Ice Age wolf remains unclear in its DNA relationships [72] to the world-wide known seven modern wolf races [72]. It seems, also those top predators became extinct in Europe with most of the "mammoth steppe/boreal forest" glacial megafauna of Europe around 24.000 BP [72], when the climate changed to colder conditions with peak in the LGM. Those last Ice Age wolves of the Sophie's Cave must have raised up their cubs in small side branches of the larger cave (see carnivore/cave bear den cave model: Fig. **12**), similar as is known for modern wolves which use cub den protecting cavities [73, 74]. Ice Age wolves also have penetrated obviously periodically into the caves for cave bear scavenging purposes, within the Late Pleistocene (see best proof in Europe in Chapter 5), whereas even modern wolves scavenge on black bears in northern America [75]. Bone growths in form of exostoses on the upper part of an ulna (Fig. **8**) might have resulted from struggles between other wolves, cave bear prey or lion/hyena antagonists, especially when taking the carnivore/cave bear den cave model into account (Fig. **12**) [62].

FINAL LATE PLEISTOCENE (AROUND 19.000 BP, MIS 2) – LAST MAXIMUM GLACIATION (LGM)

In a second phase within the early Late Late Pleistocene, the valley terrace seems to have increased a few meters in its elevation. Possibly, the colder period time frame of the MIS 2 [58] caused a final Late Pleistocene river terrace increase before the strong valley deepening erosion after the LGM [1, 12]. At this time, further sediments were flooded into the anterior halls, especially limestone gravels and till-like mixed sand/gravel/stones sediments including blocks up to 20 cm in width (Figs. **2** and **13**) [1]. Between those fluvial sediments, again large cave bear bones of *U. ingressus* were found, but in much lesser amounts, and obviously coming from the reworked bonebed layers below (see Fig. **2** and **13**). Those coarse gravels and sand sediments with glauconite content indicate higher water energies and a braided river system under a very cold (LGM-glacial) climate [1]. The glauconite might play a key role for further "glaciomodels" and studies for Upper Franconia, and is not believed to represent reworked marine Late Cretaceous glauconite (because those rocks were already eroded in the surrounding at that time, see also Chapter 2, Fig. **1**). Possibly, those glauconites originate from cold period "glaciofluvial" processes [76, 77]. The terrace elevation estimated with the relic sediments of the Sophie's Cave entrance areas (Fig. **2**) does not fit well to the model of the terraces of the neighboring Wiesent Valley, established on the stratigraphy of Zoolithen Cave [12].

Further caves have to be studied in the regions that contain similar terrace relic sediments and similar elevations, such as the Große Teufels Cave. With those two caves, Sophie's Cave and Zoolithen Cave using cave bear species/subspecies for dating within river terrace layers in caves, it can be calculated, that at the final

Late Pleistocene the Ahorn and Wisent Valleys were filled up with a minimum of four decameters of terrace sediment, and even more in the Wiesent Valley [1]. The terraces seem to have grown somehow before the LGM within the MIS 2 on an elevation of 415 meters a.s.l., and were rapidly eroded more than 40 m deep by a high energetic river system on the today's elevation of 375 meters in the Ahorn Valley within the time frame of the LGM and Holocene [1]. The sediment erosion must have happened rapidly, and under large amounts of water mass, which could be explained only with large amounts of "larger snow fields in depressions" in Upper Franconia [1, 12], which LGM glacial model is under further construction including more cave relic sediment sites. Within the past 24.000 years, the steep valley (Ahorn, Wisent, Pegnitz Valleys) morphologies of Upper Franconia below 415 m elevation developed quite recently and rapidly [1, 12].

With the LGM around 19.000 BP [10], and the inland glacier extensions from Scandinavia to Hamburg and Berlin in the North of Europe [78], and extension of the Alpine Glacier up to Munich in the South [79], or Bohemian Mountain/Bayrischer Wald (Czech Republic/Germany) smaller glaciers [80] such as the Krkonoše Mountain Chain glaciers (Czech Republic/Poland) [81], the glaciations were also present on some areas of the middle high elevated (around 1.000 m a.s.l, or even lower) regions of central Europe such as the recent discovered LGM-Brocken Mountain peak Ice Cap and valley glaciers in the Harz Mountain Range [20]. There, typical glacial deposits and geomorphological structures survived better due to the presence of hard rocks (Palaeozoic volcanites, metamorphites *etc.*), whereas in Franconia the soft dolomites weathered in glacial surface signs/morphologies much faster [1]. The Ahorn Valley is presented herein as being glaciofluvially filled during the LGM whereas a valley glacier can not be expected after classical glaciation models (Fig. **13**). There are recent kyrogen calcite descriptions from the Zoolithen Cave which were dated about 26.750 BP (= MIS 2), suggesting at minimum "permafrost conditions already few before the LGM in Upper Franconia" [82], which support the idea of latest Late Pleistocene valley "large snow fields on high elevated plateau depressions" (but no glacier caps) in the upper Wisent and branching valleys of the studied region. However, "permafrost in Upper Franconia in general" must be excluded at this elevation and still alpine boreal forest palaeoenvironment, as documented by the alpine/arctic fauna presence. There is an unpublished *Capra ibex* Linnaeus 1758 skull from the Stempfermühlen Cave (Museum Tüchersfeld near Pottenstein, Fig. **13A**), which seems to originate from "around LGM times", further supporting alpine conditions in Upper Franconia, similar as described for the Harz Mountain Range and LGM glaciation period with similar ibex and other mixed arctic/alpine faunal elements, such as wolverines [20], which latter arctic boreal forest inhabitants (*Gulo gulo, Vulpes lagopus*) are also present in the region such as in the Zoolithen Cave (Fig. **13A**) [10].

It remains unclear and is still very uncertain if smallest and limited glaciers (no ice caps, no long valley glaciers, but large snow fields in depressions or even smaller valley glaciers) might have been present during the Elsterian/Saalian but not in the Weichselian Glacials in some highest elevated (550-500 a.s.l.) areas of the Franconia Alb [1]. A snow accumulation field area situation was present upstream of the Sophie's Cave around Kirchahorn (large geomorphological depression). Even if there was no glacier ice cap around the peak of Muggendorf, at least larger amounts of wind-blown snow would have accumulated in depressions of the Jurasic plateau, forming large water masses during spring melting season. Such quick and massive water mass drainage would fit well with the braided river and river terrace development (Fig. **13**), finally causing the break-through from the Kirchahorn geomorphological depression to the Ahorn Valley. Therefore, there are LGM-glacial sedimentological signs not only in the Sophie's Cave [1], the Zoolithen Cave [10], such as undescribed yet also in the Große Teufels Cave. Dry valley sediments reaching down from the plateaus into the valleys, their valley terraces and cave sediments might result in a better understanding of the glacial landscape morphology, climate, fauna and its change around the LGM in the maximum 550 m a.s.l high elevated Upper Franconia Mountains.

REFERENCES

[1]   Diedrich C. Ice Age geomorphological Ahorn Valley and Ailsbach River terrace evolution– and its importance for the cave use possibilities by cave bears, top predators (hyenas, wolves and lions) and humans (Late Magdalénians) in the Frankonia Karst – case studies in the Sophie's Cave near Kirchahorn, Bavaria. Quat Sci J 2013; 62(2): 162-74.

[2]   Rabeder G. Neues vom Höhlenbären: Zur Morphologie der Backenzähne. Die Höhle 1983; 34(2): 67-85.

[3] Rabeder G. Die Evolution des Höhlenbärgebisses. Mitt Kommiss Quartär Österr Akad Wissensch 1999; 11: 1-102.

[4] Hofreiter M. Genetic stability and replacement in late Pleistocene cave bear populations. Abh Karst Höhlenk 2002; 34: 64-7.

[5] Rabeder G, Hofreiter M. Der neue Stammbaum der Höhlenbären. Die Höhle 2004; 55(1-4): 1-19.

[6] Stiller M, Baryshnikov G, Bocherens H, *et al.* Withering away - 25,000 years of genetic decline preceded cave bear extinction. Molec Biol Evol 2010; 27(5): 975-78.

[7] Rabeder G, Hofreiter M, Nagel D, Paabo S, Withalm G. Die neue Taxonomie der Höhlenbären. Abh Karst- Höhlenk 2002; 34: 68-9.

[8] Rabeder G, Hofreiter M, Nagel D, Whithalm G. New taxa of Alpine cave bears (Ursidae, Carnivora). Cah sci Dép Rhône-Mus Lyon 2004; 2: 49-67.

[9] Bocherens H, Stiller M, Hobson KA, *et al.* Niche partitioning between two sympatric genetically distinct cave bears (*Ursus spelaeus* and *Ursus ingressus*) and brown bear (*Ursus arctos*) from Austria: isotopic evidence from fossil bones. Quat Int 2011; 245: 249-61.

[10] Hilpert B. Studies of the morphology of the bears from the Steinberg-Höhlenruine near Hunas. Abh Naturhist Ges Nürnberg 2006; 45: 117-24.

[11] Hilpert B, Kaulich B. Eiszeitliche Bären aus der Frankenalb - Neue Ergebnisse zu den Höhlenbären aus dem Osterloch in Hegendorf, der Petershöhle bei Velden und der Gentnerhöhle bei Weidlwang. Mitt Verb dt Höhlen- Karstf 2006; 52(4): 106-13.

[12] Diedrich C. Holotype skulls, stratigraphy, bone taphonomy and excavation history in the Zoolithen Cave and new theory about Esper's "great deluge". Quat Sci J 2014; 63(1): 78-98.

[13] Diedrich C. Evolution, Horste, Taphonomie und Prädatoren der Rübeländer Höhlenbären, Harz (Norddeutschland). Mitt Verb dt Höhlen-Karstf 2013; 59(1): 4-29.

[14] Tsoukala E, Chatzopoulou K, Rabeder G, Pappa S, Nagel D, Withalm G. Paleontological and stratigraphical research in Loutra Ariedas Bear Cave (Almopia Speleopark, Pella, Macedonia, Greece). Scie Ann School Geol Arist Univ Thessaloniki (AUTH) 2006; spec vol 98: 41-67.

[15] Diedrich C. Extinctions of Late Ice Age cave bears as a result of climate/habitat change and large carnivore lion/hyena/wolf predation stress in Europe. ISRN Zoology 2013; 1-25.

[16] Münzel SC, Stiller M, Hofreiter M, Mittnik A, Conard NJ, Bocherens H, Pleistocene bears in the Swabian Jura (Germany): Genetic replacement, ecological displacement, extinctions and survival. Quat Int 2011; 245: 225-37.

[17] Rosenmüller JC. Die Merkwürdigkeiten der Gegend um Muggendorf. Unger 1804; p. 90.

[18] Diedrich C. The rediscovered cave bear "*Ursus spelaeus* Rosenmüller 1794" holotype of the Zoolithen Cave (Germany) from the historic Rosenmüller collection. Acta Carsol Slov 2009; 47(1): 25-32.

[19] Clark PU, Dyke A, Shakun JD, *et al.* The Last Glacial Maximum. Science 2009; 325(5941): 710-4.

[20] Diedrich C. Impact of the German Harz Mountain Weichselian ice-shield and valley glacier development onto Palaeolithics and megafauna disappearance. Quat Sci Rev 2013; 82: 167-98.

[21] Diedrich C. Cave bear killers and scavengers from the last Ice Age of central Europe: Feeding specializations in response to the absence of mammoth steppe fauna from mountainous regions. Quat Int 2011; 255: 59-78.

[22] Diedrich C. The Late Pleistocene spotted hyena *Crocuta crocuta spelaea* (Goldfuss 1823) population with its type specimens from the Zoolithen Cave at Gaillenreuth (Bavaria, South Germany) – a hyena cub raising den of specialized cave bear scavengers in boreal forest environments of Central Europe. Hist Biol 2011: 1-33.

[23] Diedrich C. Cave bear killers, scavengers between the Scandinavian and Alpine Ice shields – the last hyenas and cave bears in antagonism – and the reason why cave bears hibernated deeply in caves. Stalactite 2009; 58(2): 53-63.

[24] Diedrich C. The largest European lion *Panthera leo spelaea* (Goldfuss) population from the Zoolithen Cave, Germany – specialized cave bear predators of Europe. Hist Biol 2011; 23(2-3): 271-311.

[25] Bocherens H, Drucker DG, Bonjean D, *et al.* Isotopic evidence for dietary ecology of cave lion (*Panthera spelaea*) in North-Western Europe: Prey choice, competition and implications for extinction. Quat Int 2011; 245: 249-61.

[26] Diedrich C. Oldest Late Palaeolithic cave bear hunters in Europe (Hermanns's Cave, northern Germany) the end of the Ice Age. (in review).

[27] Münzel SC, Conrad N. Cave bear hunting in the Hohle Fels, a cave site in the Ach Valley, Swabian Jura. Rev Paléobiol 2004; 23: 877-85.

[28] Goldfuss GA. Die Umgebungen von Muggendorf. Ein Taschenbuch für Freunde der Natur und Alterthumskunde. Erlangen 1810; p. 351.

[29] Diedrich C. The rediscovered holotypes of the Upper Pleistocene spotted hyena *Crocuta crocuta spelaea* (Goldfuss 1823) and the steppe lion *Panthera leo spelaea* (Goldfuss 1810) and taphonomic discussion to the Zoolithen Cave hyena den at Geilenreuth (Bavaria, South-Germany). Zool J Linn Soc London 2008; 154: 822-31.

[30] Hemmer H. Pleistozäne Katzen Europas – eine Übersicht. Cranium 2003; 20: 6-22.

[31] Diedrich C. Steppe lion *Panthera leo spelaea* (Goldfuss 1810) remains imported by Ice Age spotted hyenas *Crocuta crocuta spelaea* (Goldfuss 1823) from the Perick Caves, a Late Pleistocene hyena den in Northern Germany. Quat Res 2009; 71(3): 361-74.

[32] Diedrich C. Late Pleistocene steppe lion *Panthera leo spelaea* (Goldfuss 1810) footprints and bone remains from open air sites in northern Germany – evidence of hyena-lion antagonism in Europe. Quat Sci Rev 2011; 30: 1883-906.

[33] Diedrich C. The Late Pleistocene *Panthera leo spelaea* (Goldfuss 1810) skeletons from the Sloup and Srbsko Caves in Czech Republic (central Europe) and contribution to steppe lion cranial pathologies and postmortally damages as results of interspecies fights, hyena antagonism and cave bear attacks. Bull Geoscie 2011; 86(4): 817-40.

[34] Diedrich C. Late Pleistocene Eemian hyena and steppe lion feeding strategies on their largest prey - *Palaeoloxodon antiquus* Falconer and Cautley 1845 at the straight-tusked elephant graveyard and Neanderthal site Neumark-Nord Lake 1, Central Germany. Archaeol Anthropol Sci 2014; 6(3): 271-91.

[35] Diedrich C. Pleistocene *Panthera leo spelaea* (Goldfuss 1810) remains from the Balve Cave (NW Germany) – a cave bear, hyena den and Middle Palaeolithic human cave, and review of the Sauerland Karst lion sites. Quaternaire 2011; 22(2): 105-27.

[36] Schaller G. The Serengeti Lion. A Study of Predator-Prey Relations. The University of Chicago Press, Chicago 1972; p. 494.

[37] Diedrich C. Ice Age spotted hyena feeding behavior on Neanderthals in Europe – impact on burial destructions. Chron Sci 2014; 1(1): 1-34.

[38] Bosinski G. Die große Zeit der Eiszeitjäger. Europa zwischen 40.000 und 10.000 v. Chr. Jb Röm-Germ Zentralmus Mainz 1987; 34: 131-9.

[39] Diedrich C. Ichnological and ethological studies in one of Europe's famous bear den in the Urşilor Cave (Carpathians, Romania). Ichnos 2011; 18(1): 9-26.

[40] Rothschild BM, Diedrich C. Comparison of pathologies in the extinct Pleistocene Eurasian steppe lion *Pantherea leo spelaea* (Goldfuss 1810) to those in the modern lion, *Panthera leo* – Results of fights with hyenas, bears and lions and other ecological stress. Int J Paleopath 2012; 2: 187-98.

[41] Staatsarchiv Bayreuth, unpublished documents K3FVVIII/321. Regierung des Obermain-kreises/Oberfranken; 1833.

[42] Goldfuss GA. Osteologische Beiträge zur Kenntnis verschiedener Säugethiere der Vorwelt. VI. Ueber die Hölen-Hyäne (*Hyaena spelaea*). Nov Act Phys-Med Acad Caes Leopold-Carol Nat Curios 1823; 3(2): 456-90.

[43] Buckland W. Reliquiae Diluvianae, or observations on the organic remains contained in caves, fissures, and diluvial gravel, and other geological phenomena, attesting the action of an universal deluge. London: J. Murray 1823; p. 303.

[44] Diedrich C. The Ice Age spotted *Crocuta crocuta spelaea* (Goldfuss 1823) population, their excrements and prey from the Late Pleistocene hyena den Sloup Cave in the Moravian Karst; Czech Republic. Hist Biol 2012; 24(2): 161-85.

[45] Diedrich C. Europe's first Upper Pleistocene *Crocuta crocuta spelaea* (Goldfuss 1823) skeleton from the Koněprusy Caves - a hyena cave prey depot site in the Bohemian Karst (Czech Republic) – Late Pleistocene woolly rhinoceros scavengers. Hist Biol 2012; 24(1): 63-89.

[46] Diedrich C, Žák, K. Prey deposits and den sites of the Upper Pleistocene hyena *Crocuta crocuta spelaea* (Goldfuss 1823) in horizontal and vertical caves of the Bohemian Karst (Czech Republic). Bull Geosci 2006; 81(4): 237-76.

[47] Diedrich C. Periodical use of the Balve Cave (NW Germany) as a Late Pleistocene *Crocuta crocuta spelaea* (Goldfuss 1823) den: Hyena occupations and bone accumulations vs. Human Middle Palaeolithic activity. Quat Int 2011; 233: 171-84.

[48] Diedrich C. Late Pleistocene spotted hyena den sites and specialized rhinoceros scavengers in the Thuringian Mountain Zechstein karst (Central Germany). Quat Sci J 2015; 64(1): 29-45.

[49] Fosse P, Brugal JP, Guadelli JL, Michel P, Tournepiche JF. Les repaires d' hyenes des cavernes en Europe occidentale: presentation et comparisons de quelques assemblages osseux. In: Economie Prehistorique, Les comportements de substance au Paleolithique, XVIII Rencontres internationales d' Archeologie et d' Historie d' Antibes, Editions APDCA, Sophia Antipolis 1998; pp. 44-61.

[50] Musil R. Die Höhle "Sveduv stůl", ein typischer Höhlenhyänenhorst. Anthropos NS 1962; 5(13): 97-260.

[51] Ehrenberg K, Sickenberg O, Stifft-Gottlieb A. Die Fuchs- oder Teufelslucken bei Eggenburg, Niederdonau. 1 Teil. Abh Zool-Bot Ges 1938; 17(1): 1-130.

[52] Diedrich C. Typology of Ice Age spotted hyena *Crocuta crocuta spelaea* (Goldfuss 1823) coprolite aggregate pellets from the European Late Pleistocene and their significance at dens and scavenging sites. New Mex Mus Nat Hist Sci Bull 2012; 57: 369-77.

[53] Ziegler R. Das Mammut (*Mammuthus primigenius* Blumenbach) von Siegsdorf bei Traunstein (Bayern) und seine Begleitfauna. Münchner geowiss Abh A Geol Paläont 1994; 26: 49-80.

[54] Siegfried P. Das Mammut von Ahlen. Paläont Z 1959; 33: 172-84.

[55] Walker MJC, Björck S, Lowe JJ, *et al.* Isotopic events in the GRIP ice core: a stratotype for the Late Pleistocene. Quat Sci Rev 1999; 18: 1143-50.

[56] Diedrich C. Mammoth scavengers in Europe - the Ice Age spotted hyenas and steppe lions and their feeding strategies on their largest prey (in review).

[57] Diedrich C. Von eiszeitlichen Fleckenhyänen benagte *Mammuthus primigenius* (Blumenbach 1799) - Knochen und -Knabbersticks aus dem oberpleistozänen Perick-Höhlenhorst (Sauerland) und Beitrag zur Taphonomie von Mammutkadavern. Philippia 2005; 12(1): 63-84.

[58] White P, Diedrich C. Taphonomy story of a modern African elephant *Loxodonta africana* carcass on a lakeshore in Zambia (Africa). Quat Int 2012; 276/277: 287-96.

[59] Diedrich C. The Late Pleistocene *Crocuta crocuta spelaea* (Goldfuss 1823) population from the Emscher River terrace hyena open air den Bottrop and other sites in NW-Germany – woolly rhinoceros scavengers and their bone accumulations along rivers in lowland mammoth steppe environments. Quat Int 2012; 276-277: 93-119.

[60] Kruuk H, The spotted hyena. A story of predation and social behavior. The University of Chicago Press, Chicago 1972; p. 352.

[61] Sutcliffe AJ. Spotted Hyaena: crusher, gnawer, digester and collector of bones. Nature 1970; 227: 110-3.

[62] Diedrich C. Palaeopopulations of Late Pleistocene top predators in Europe: Ice Age spotted hyenas and steppe lions in battle and competition about prey. Paleont J 2014; 1-34.

[63] Diedrich C. Eingeschleppte und benagte Knochenreste von *Coelodonta antiquitatis* (Blumenbach 1807) aus dem oberpleistozänen Fleckenhyänenhorst Perick-Höhlen im Nordsauerland (NW Deutschland) und Beitrag zur Taphonomie von Wollnashornknochen in Westfalen. Mitt Höhlen-Karstf 2008; 4: 100-17.

[64] Diedrich C. Specialized horse killers in Europe – foetal horse remains in the Late Pleistocene Srbsko Chlum-Komín Cave hyena den in the Bohemian Karst (Czech Republic) and actualistic comparisons to modern African spotted hyenas as zebra hunters. Quat Int 2010; 220(1-2): 174-87.

[65] Diedrich C. Late Pleistocene *Crocuta crocuta spelaea* (Goldfuss 1823) clans as prezewalski horse hunters and woolly rhinoceros scavengers at the open air commuting den and contemporary Neanderthal camp site Westeregeln (central Germany). J Archaeol Sci 2012; 39(6): 1749-67.

[66] Stiner MC. Comparative ecology and taphonomy of spotted hyenas, humans, and wolves in Pleistocene Italy. Rev Paléobiol 2004; 23(2): 771-85.

[67] Diedrich C. "Neanderthal bone flutes" – simply products of Ice Age spotted hyena scavenging activities on cave bear cubs in European cave bear dens. Roy Soc Open Sci 2015; 2: 14002.

[68] Turk M, Dimakaroski L. Neanderthal flute from Divje Babe I: old and new findings. Op Inst Archaeol Slov 2011; 21: 251-64.

[69] Diedrich C. Cracking and nibbling marks as indicators for the Upper Pleistocene spotted hyaena as a scavenger of cave bear (*Ursus spelaeus* Rosenmüller 1794) carcasses in the Perick Caves den of Northwest Germany. Abh Naturhist Ges Nürnberg 2005; 45: 73-90.

[70] Reynolds SH. A monograph of the British Pleistocene Mammalia. The Canidae. Palaeontogr Soc Monogr 1909; 2(3): 1-28.

[71] Diedrich C. The largest European Late Pleistocene wolf population from the Zoolithen Cave (Bavaria, Germany), taxonomy, sexual dimorphism – and taphonomic contribution to the extinct European *Canis lupus spelaeus* (Goldfuss 1823) as cave bear scavengers. (in prep).

[72] Leonard JA, Carles Vilà C, Dobbs KF, Koch PL, Wayne RK, Van Valkenburgh B. Megafaunal extinctions and the disappearance of a specialized wolf ecomorph. Curr Biol 2007; 17(13): 1146-50.

[73] Bibikow DI. Der Wolf - Canis lupus. Wittenberg, Neue Brehm-Bücherei 2003; p. 198.

[74] Mech LD. The Arctic Wolf: Living with the Pack. Stillwater: Voyageur Press Inc 1988: p. 128.

[75] Rogers LL, Mech D. Interactions of wolves and black bears in northeastern Minnesota. J Mammal 1981; 62(2): 434-6.

[76] Skiba M. Clay mineral formation during podzolization in an alpine environment of the Tatra Mountains, Poland. Clay Miner 2007; 55: 618-4.

[77] Williams GE. Subglacial meltwater channels and glaciofluvial deposits in the Kimberley Basin, Western Australia: 1.8 Ga low-latitude glaciation coeval with continental assembly. J Geol Soc 2004; 162(1): 111-24.

[78] Weymann H-J, Feldmann L, Bombien H. Das Pleistozän des nördlichen Harzvorlands – eine Zusammenfassung. Eiszeit Gegenw 2005; 55: 43-63.

[79] Heuberger H. Die Alpengletscher im Spät- und Postglazial. Eine chronologische Übersicht. Eiszeit Gegenw 1968; 19: 270-5.

[80] Mentlík P, Minár J, Břízová E, Lisa´ L, Táborík P, Stacke V. Glaciation in the surroundings of Prášilské Lake (Bohemian Forest, Czech Republic). Geomorphology 2010; 117(1/2): 181-94.

[81] Engel Z, Nyvlt D, Křižek M, Treml V, Jankovská V, Lisá L. Sedimentary evidence of landscape and climate history since the end of MIS3 in the Krkonoše Mts., Czech Republic. Quat Sci Rev 2010; 29: 913-27.

[82] Richter DK, Harder M, Niedermayr A, Scholz D. Zopfsinter in der Zoolithenhöhle: Erstfund kyrogener Calcite in der fränkischen Alb. Mitt Verb dt Höhlen-Karstf 2014; 60(2): 36-41.

CHAPTER 7

LATE PLEISTOCENE ARCHAEOLOGY

Abstract: The Sophie's Cave was used during the last dry cave stage just before the maximum glaciation by the first modern humans, Late Palaeolithic Early Gravettian reindeer hunters, but not as short-term hunting camp or or settlement, but as shamanic sanctuary. Those hunters seem to have deposited year after year selected larger male shed reindeer antlers (one C^{14}-dated 30.830-30.340 cal. BP) in only one of the deeper chambers of the cave, possibly also two mammoth pelvic halves (one C^{14}-dated 29.340-28.600 cal. BP) and other bones of the megafauna. Already before the Last Maximum Glaciation (= LGM, around 19.000 BP) humans and the cave bear boreal fauna disappeared in the Franconia Karst region, similar as all over Central Europe. The valley was resettled by the last reindeer hunters at the end of the glaciation by Epipalaeolithics, which left remains in other smaller cave entrances or rock shelters in lower elevations along the Wisent and Ahorn valleys.

Keywords: Little presence of Neanderthals, Late Palaeolithic Early Gravettian reindeer hunters, reindeer antler and bone depot, shamanic cave use of Sophie's Cave, Epipal-aeolithics.

NEANDERTHALS IN UPPER FRANCONIA

At the same time of the cave den use by bears, and few hyenas and wolves of the Sophie's Cave and other caves in Upper Franconia (Fig. **1C**), Neanderthal humans of the Middle Paleolithic times used only a single small cavity in higher elevated rock shelter of the Hasenloch Cave near Pottenstein [1, 1C] whereas the presence of a single tooth from the cave bear den Hunas Cave ruin of Franconia [2] remains unresolved in the taphonomy, because no artifacts have been found with this single human remain. The Neanderthal cranial remain was possibly imported by hyenas, as known for several cave sites in Europe now, where Neanderthals were consumed and partly imported by carcass body parts to their cave dens [3]. Only in the Hasenloch Cave, stone tool and flake artifacts such as porcupine and other bone remains have been described [1, 4]. To date, it is unproven for Europe that Neanderthals killed and consumed the flesh of cave bears [5 - 7]. The "Nean-

Cajus G. Diedrich

derthal caused holes in two skulls" found in the Zoolithen Cave were misidentified by Groiss [8] to have resulted from "spears", indeed, those are described with even more crania to represent non-healed wounds of battles between other cave bears or top predators (lions, hyenas) [9]. A cave bear hunt has not yet been shownin Upper Fanconia in any cave bear den cave.

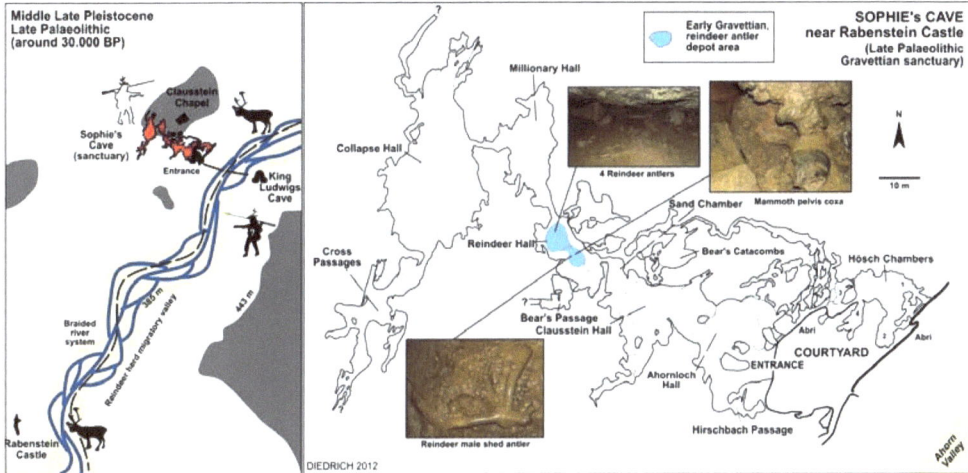

Figure 1. Reindeer antlers in the Reindeer Hall which were accumulated by Late Palaeolithic Early Gravettian humans. Most of the antlers which have been collected and removed are smashed from the speleothem layer, in historic times, with mainly fragments being left. One nearly complete shed antler dated (C^{14} age 30.830-30.340 cal. BP) is in the center of the Hall, opposite the mammoth coxa (C^{14} age 29.340-28.600 cal. BP, cf. Chapter 6, Fig. **9**) (coll. Rabenstein Castle Museum, some *in situ*).

FIRST CAVE BEAR HUNTERS - LATE PALAEOLITHIC EARLY GRAVETTIANS (AROUND 30.000 BP)

The first secure proof of cave bear hunting by modern early Late Paleolithic Aurignacians to Gravettian humans is from the Hohle Fels Cave near Schelklingen of the Swabian Alb region [7]. There, a thoracic vertebra of an approximate 30.000 BP dated (Late Aurignacian) adult cave bear (possibly = *U. ingressus*) contains a projectile fragment [7]. Newest discoveries in the Herman's Cave in the Harz Mountains of northern Germany indicate also Aurignacien propulsor weapon hunting technique and even cave bear butchering (cut marks on cave bear bones) deeper in caves [6]. There are now some caves in Europe from Aurignacien-Gravettian times, where the cave bear hunt becomes obvious in boreal forest cave-rich and cave bear population regions [6]. With this knowledge, the Solutréen propulsor spear point from the Große Teufels Cave, Upper Franconia [1, 4] becomes of interest. This was also found within a larger cave bear den (Fig. **1**) entrance area, and is a single find without other artifacts. Whereas in several caves of Aurignacian and Gravettian ages cave bear hunt must have taken at several European caves [6, 7], cave bear "hunting" by humans or cave bear "cult" by humans can not be observed in the Sophie's Cave, or the surrounding Upper Franconia caves, yet. Most of the sites were destroyed or partly damaged by pickaxe shovel excavations already historically in many parts and were modified especially by spelunkers and cave visitors since the 19th century. The past research did not focus on such questions about "bone taphonomy" or any bone research at all (*e.g.* Große Teufels Cave). However, there is an obvious Late Palaeolithic (Early Gravettian) non-camp site cave use of the Sophie's Cave as described below.

Early Gravettian Shamanic Sanctuary Around 30.000 BP

In the Reindeer Hall, calculated by the historical descriptions [10, 11] and the new research and documentation of all visible antlers and new discovered fragments (Fig. **1**) approximately 100 antlers must have been present there before 1833.

After descriptions and new finds/observations, all, or most of them, must have been antler sheds. One large shed male antler (Chapter 7 Fig. **2** or Fig. **2.1**) was dated with an C^{14} age of 30.830 - 30.340 cal. BP (in the laboratory of Beta Analytic Inc., USA, no. SOPHIER-EN001). This age falls into the Early Gravettian period compared to Aurignacian/ Gravettian ages *e.g.* of the southern German Geißenklösterle Cave [12]. Calcuated from the preserved remains, all of them are from large, adult male individuals (Chapter 7 Fig. **2**, Fig. **2**) with diameters of 3-4 cm (females and calves are around 2 cm [13]). From the historical descriptions, there was also a skull with attached antlers [11] and nearby, today's most complete antler (Chapter 3 Fig. 47). Most preserved specimens were left *in situ* under a large ceiling block niche in the lower part of the Reindeer Hall (Chapter 7 Fig. **1-2**).

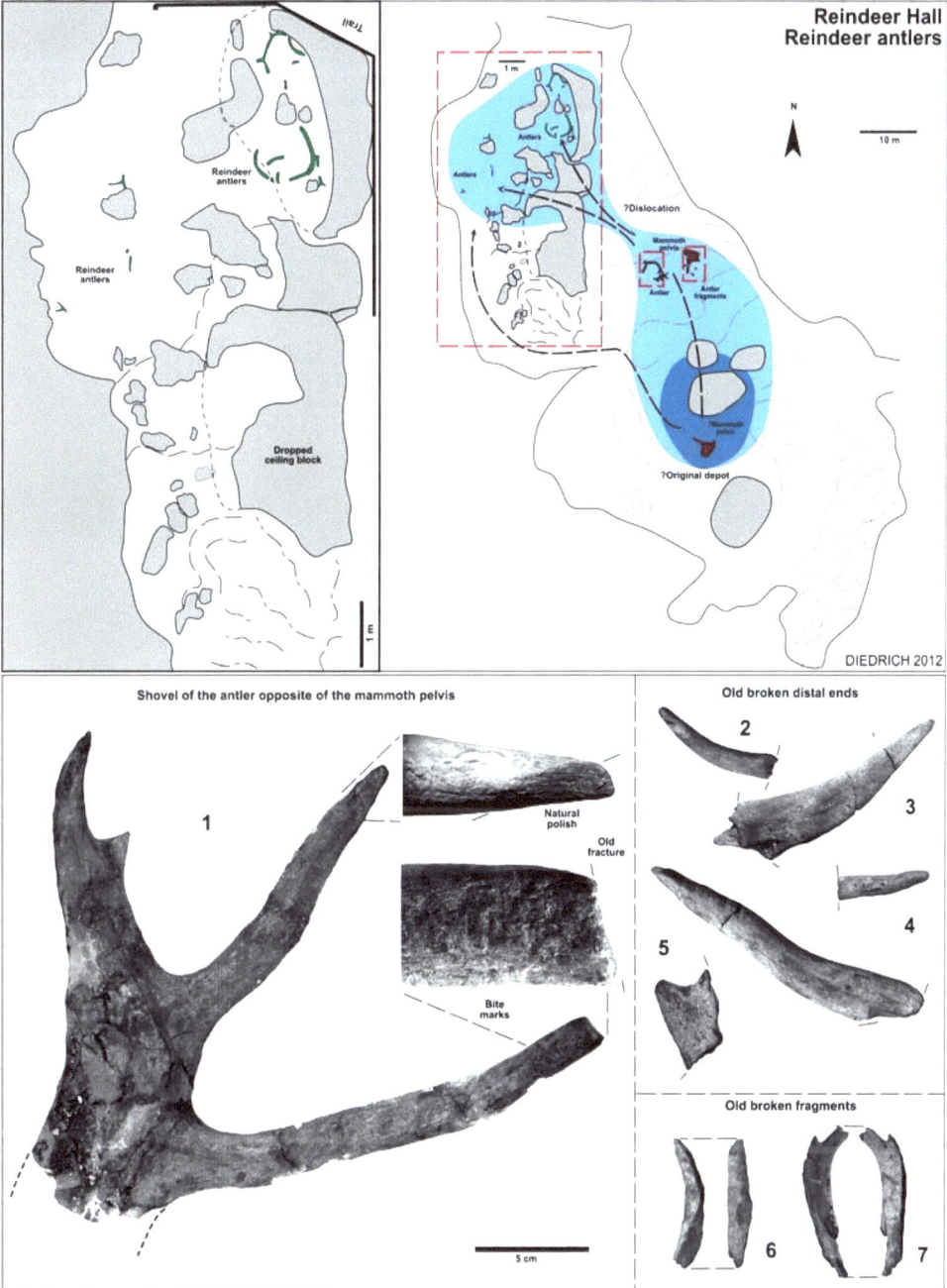

DIEDRICH 2012

Shovel of the antler opposite of the mammoth pelvis

Old broken distal ends

Old broken fragments

(Fig 2) contd.....

Figure 2. Above: Reindeer antler remains (most or all shed antlers selected from large, strong male reindeer) which were found only in the Reindeer Hall (blue area) and possibly deposited schematically by Late Palaeolithic (Gravettian) reindeer hunters (coll. Rabenstein Castle Museum, some *in situ*). 1. Distal antler part of the nearly complete shed antler (cf. Chapter 6, Fig. **9**) being ^{14}C-dated 30.830-30.340 cal. BP. 2-5. Antler fragments with old natural breakages. 6-7. Antler fragments. Below: Photo into the Reindeer Hall.

Whether the Ice Age spotted hyenas [14], wolves [15 - 19], nor wolverines [16, 17] could have "selected" purposely only male shed antlers, or deposited them in such a high amount and in a limited area within the upper Reindeer Hall – this is securely and obviously a result only of human activity.

Even if some of the new excavated antler ends have small bite impact marks (Fig. **1**), which seem to be the result of wolves/wolverines, they seem to have been chewed already outside the cave (similar as hyena/lion chew damaged mammoth coxa), before human collecting and importation. Such chewing is even today common in Scandinavian/ Canadian wolves/wolverines and other animals [13, 15 - 19]. The diagonal polishing on antler ends (Fig. **1**) are not of human origin and happened naturally [13], and are common, especially on male antlers. Further-more at hyena dens typical "chewed antler bases" are found only, because these carnivores leave only in few amounts (up to max. 5-10 pieces) with massive bite damage on the first 10-20 cm of the "rose part" of the antlers [14].

The only ones which might have accumulated the large amounts of antlers in the Sophie's Cave Reindeeer Hall are the Late Palaeolithic "reindeer hunters" of the final Ice Age, who hunted and depended mainly on reindeer and horses as main game [20 - 21]. Possibly, in southern Germany, and other regions of northern Germany (cave-rich boreal forest regions), those reindeer hunters did not paint animals in caves, as is common in SW-Europe (France, Spain, *e.g.* Altamira-, Chauvet-, Lascaux Caves [20 - 23]). They seem to have practiced at least parallel a different cult by choosing deeper cave areas, accumulating reindeer antlers over generations for shamanic rituals (as known for cave art [20 - 23]). The "male" antlers of the Sophie's Cave would have represented the "power of the big bucks". Similar reindeer antler deposition acccumulations were found in northern German caves (Oeger Cave [24] and undescribed in several others) and on a new open air site, Westeregeln in central Germany, whereas the latter is unclear in the age attribution to Neanderthals or Late Palaeolithics [25].

Figure 3. Possible situatuion of a shamanic deposition of reindeer antlers only in the Reindeer Hall of the Sophie's Cave by Late Palaeolithic Early Gravettian humans (one antler is C[14]-dated 30.830-30.340 cal. BP).

In the Sophie's Cave, the Early Gravettian shamans possibly imported every year a couple of antlers, and possibly even arranged a reindeer skull on a stick (Fig. **3**), which is also known from a northern German final Late Palaeolithic site [26]. Wheter the two mammoth pelvic "halves" (coxae) are also imported by those humans remains unclear. Possibly those reflect the female fertility (pelvic – birth canal) in a shamanic ritual. It seems, the coxae were at least human imported and dumped with the other bones below the main stalagmite/stalagmites below the "Bienenkorb" in the Reindeer Hall. In hyena den caves the situation is much different. Such carnivore imported mammoth pelvic remains are much more fragmented smaller and have chew/bite marks nearly all around, especially in the soft spongiosa at communal and birth dens [27]. It seems here, opposite, after import by humans, few were chewed by carnivores within the cave (see also chewed bones of cave bears), which is indicated by a fragment found besides the pelvic, that was clearly chewed off (Chapter 6 Fig. **10**). The selection of the male antlers by humans does fit to known"reindeer fertility ceremonies" of natural hunter groups [28]. The hunters expected with those rituals possibly new and abundant reindeer herds in the next season, similar as practiced in the Medieval and historic times of "reindeer breeders" in Scandinavia or Siberia, the Sami [29 - 32]. These Scandinavian Sami people accumulated due to religious shamanic purposes, reindeer antlers as well as skulls from wolves and brown bears into the 16th and 17th centuries [29 - 32].

Also from Siberia those types of cultic places, where bones and antlers were accumulated for shamanic reasons, are figured from the Waigatsch Islands [29]. In Scandinavia small cavity niches or small caves (large caves are absent) were used for the large or small deposits [29 - 32].

At those cultic places, and often on overview points, year by year collected reindeer antlers and lesser skulls and postcranial bones were placed, with these deposits growing in "bone amounts" over generations [29 - 32]. This similar reindeer antler deposition at least was initially discussed with the northern German Oeger Cave (Sauerland Karst, Fig. **4**) where Late Palaeolithic Late Magdalénians deposited hundreds of reindeer antlers. There, different to the Sophie's Cave, only female/calf antlers were accumulated [24].

This new initial model presented herein of "reindeer antler depots – versus cave art - with mainly accumulated selected shed antlers fits well with the Sophie's Cave bone taphonomy and dating. This would even further explain the absence of engravings and cave art (Aurignacien to Solutréen) in Upper Franconian and other northern German caves (Fig. 4) as well, whereas the next very few cave "rock art " (no paintings) was reported from the Early Gravettian (there nearly similar aged with C^{14}-AMS about 29.000 BP) of the southwestern German Geißenklösterle Cave (Fig. 4) [12].

During the Late Palaeolithic times, it was possible to enter the Sophie's Cave Reindeer Hall through a small passage in the Sand Chamber. This was much later closed about some thousands of years (Post-LGM = after 19.000 BP). The Gravettian reindeer hunters were able to see the mostly formed "Elefantenohr- and Bienenkorb" stalagmites and stalactites, and other speleothems in that first hidden hall, which may have been inspiration for the depositing of the antlers/bones nearby and below those in the upper part of the hall (Figs. 1-2) [33]. The antlers were much later, some thousands of years after the Gravettian, redeposited mainly by dripping and fluent water and gravity within the humid and climate changing Dryas-Alleröd interstadial times (about 16.000-14.000 BP).

Those antlers seem to have been smashed naturally (no modification by humans for tool production). This correlates to the main ceiling collapses period. Massive blocks falling would have easily smashed antlers, which would best explain several isolated, old-broken distal ends and fragments which spread within the hall (Fig. 1). Those two non-anthropogen taphonomic reasons would explain why they are today scattered, especially down toward the Reindeer Hall, with several antlers that survived below a very large block (Fig. 1). Finally, in this period of main speleothem genesis, all the antlers were encrusted and almost completely covered by the youngest speleothem layer (Chapter 3 Fig. 47B) [33].

Newest engravings (female symbols) of the final Late Palaeolithic in the Upper Franconia Mäander Cave [34] demostrate the presence of Late Magdalénians (about 16.000-14.000 BP) several thousands of years after the Last Glacial Maximum (around 19.000 BP) in the Upper Franconia region and indicate a different shamanic use of caves (no cave art paintings, and rarely engravings on speleothems [33]). Epiplalaeolithic reindeer hunters are recorded to have used a small cave/rock shelter within the few kilometers south of Sophie's Cave in the Ahorn Valley at the Rennerfels rock shelter where stone and bone tools have been excavated, unsystematically, in historic times [35, 36].

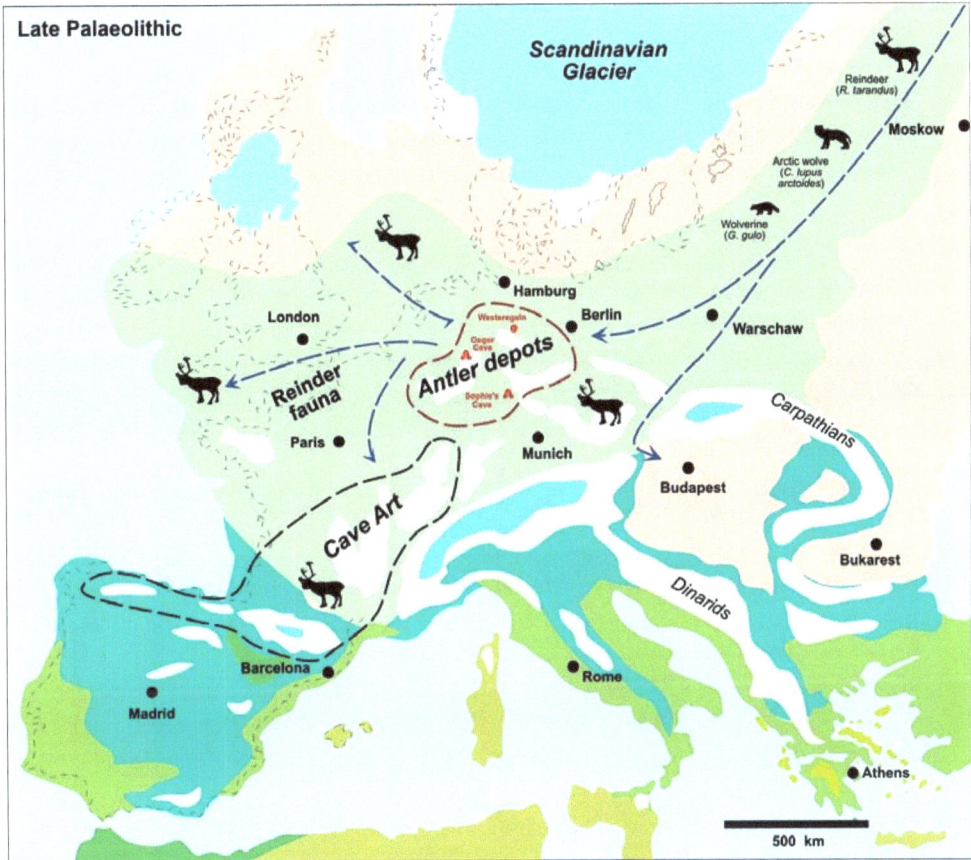

Figure 4. Model of differences in the shamanic cave use by Late Palaeolithic Cromagnon humans in Europe. Whereas in the Southwest the cave art was continous between the Aurignacien-Magdalénian, this was absent in Central Europe (*e.g.* all German caves). There, a very few Early Gravettian rock art, Late Magdalénian engravings are known. Instead, in those regions antler/bone deposits seem to have been made in the boreal forest regions mainly in caves, possibly also at open air sites (cave art region composed after [20 - 23].

With the end of the Ice Age, some thousands of years after the peak of the LGM, the reindeer herds disappeared at the beginning of the Holocene also due to climate change from southern and central Europe and migrated to the North (Canada, Scandinavia, Siberia) [13], which finally caused the disappearance of "human reindeer hunter" gatherers in Central Europa all over [20], such as in Upper Franconia [4].

REFERENCES

[1] Ranke J. Das Zwergloch und Hasenloch bei Pottenstein in Oberfranken. Beitr Anthrop Urgesch Bayerns 1879; 2: 209-10.

[2] Hilpert B. Studies of the morphology of the bears from the Steinberg-Höhlenruine near Hunas. Abh Naturhist Ges Nürnberg 2006; 45: 117-24.

[3] Diedrich C. Ice Age spotted hyenas as Neanderthal exhumers and scavengers in Europe. Chronicles Sci 2014; 1(1): 1-34.

[4] Diedrich C. Ice Age geomorphological Ahorn Valley and Ailsbach River terrace evolution– and its importance for the cave use possibilities by cave bears, top predators (hyenas, wolves and lions) and humans (Late Magdalénians) in the Frankonia Karst – case studies in the Sophie's Cave near Kirchahorn, Bavaria. Quat Sci J 2013; 62(2): 162-74.

[5] Auguste P. La chasse à l'ours au Paléolithique moyen: mythes, réalités et état de la question. Brit Archaeol Rep Int Ser 2003; 1105: 135-42.

[6] Diedrich C. Oldest Late Palaeolithic cave bear hunters in Europe (Hermanns's Cave, northern Germany) the end of the Ice Age. (in review).

[7] Münzel SC, Conrad N. Cave bear hunting in the Hohle Fels, a cave site in the Ach Valley, Swabian Jura. Rev Paléobiol 2004; 23: 877-85.

[8] Groiss JT. Über pathologische Bildungen an Skelettresten jungquartärer Säugetiere aus der Zoolithenhöhle bei Burggaillenreuth. Geol Bl NO-Bayern 1978; 28(1): 1-21.

[9] Diedrich C. Holotype skulls, stratigraphy, bone taphonomy and excavation history in the Zoolithen Cave and new theory about Esper's "great deluge". Quat Sci J 2014; 63(1): 78-98.

[10] Sternberg K. Vortrag des Präsidenten Grafen Kaspar Sternberg in der allgemeinen Versammlung des böhmischen Museums in Prag. Verh Ges vaterl Mus Böhmen Prag 1835; 12-30.

[11] Staatsarchiv Bayreuth, unpublished documents K3FVVIII/321. Regierung des Obermainkreises/Oberfranken 1837; p. 44.

[12] Richter D, Waiblinger J, Rink WJ, Wagner GA. Thermoluminescence, electron spin resonance and C^{14}-dating of the Late Middle and Early Palaeolithic site of Geißenklösterle Cave in Southern Germany. J Archaeol Sci 2000; 27: 71-89.

[13] Weinstock J. Late Pleistocene reindeer populations in Middle and Western Europe. BioArchaeologica 2000; 3: 1-307.

[14] Diedrich C. Palaeopopulations of Late Pleistocene top predators in Europe: Ice Age spotted hyenas and steppe lions in battle and competition about prey. Paleont J 2014; 1-34.

[15] Diedrich C. Extinctions of late ice age cave bears as a result of climate/habitat change and large carnivore lion/hyena/wolf predation stress in Europe. ISRN Zoology 2013; 1-25.

[16] Diedrich C. Evolution, Horste, Taphonomie und Prädatoren der Rübeländer Höhlenbären, Harz (Norddeutschland). Mitt Verb dt Höhlen-Karstf 2013; 59(1): 4-29.

[17] Diedrich C, Copeland J. Upper Pleistocene *Gulo gulo* (Linné 1758) remains from the Srbsko Chlum-Komin hyena den cave in the Bohemian Karst, Czech Republic, with comparisons to contemporary wolverines. J Cave Karst Stud Am 2009; 72(2): 1222-7.

[18] Mech LD, Packard JM. Possible use of (*Canis lupus*) den over several centuries. Canadian Field Naturalist 1990; 104: 484-5.

[19] Mech LD The Arctic Wolf: Living with the Pack. Voyageur Press Inc.: Stillwater 1988; p. 128.

[20] Bosinski G. Die große Zeit der Eiszeitjäger. Europa zwischen 40.000 und 10.000 v. Chr. Jb Röm-Germ Zentralmus Mainz 1987; 34: 131-9.

[21] Lumley H, Couraud C, Delloc B, *et al.* Art et civilisations des chasseurs de la préhistore. 34000-8000 ans av. J.-C. Laboratoire de Préhistoire du Muséum National d'Histoire Naturelle Musée de l'Homme. Paris, Imprimerie Louis-Jean 1984; p. 415.

[22] Braem H. Die magische Welt der Schamanen und Höhlenmaler. Köln, Dumont-Buchverlag 1994; p. 276.

[23] Chauvet J-M, Deschamps BE, Hillaire C. Grotte Chauvet. Altsteinzeitliche Höhlenkunst im Tal der Ardèche. Sigmaringen, Thorbecke Speläo 1 1995; p. 120.

[24] Bleicher W. Die Oeger-Höhle – eine Kultstätte altsteinzeitlicher Rentierjägergruppen. Hohen-limburger Heimatbl 1993; 9: 309-23.

[25] Diedrich C. Impact of the German Harz Mountain Weichselian ice-shield and valley glacier development onto Palaeolithics and megafauna disappearance. Quat Sci Rev 2013; 82: 167-98.

[26] Tromnau G. Den Rentierjägern auf der Spur - 50 Jahre Eiszeitforschung im Ahrensburger Tunneltal. Neumünster, Karl-Wachholtz Verlag 1980; p. 46.

[27] Diedrich C. Von eiszeitlichen Fleckenhyänen benagte *Mammuthus primigenius* (Blumenbach 1799)-Knochen und -Knabbersticks aus dem oberpleistozänen Perick-Höhlenhorst (Sauerland) und Beitrag zur Taphonomie von Mammutkadavern. Philippia 2005; 12(1): 63-84.

[28] Baales M. Umwelt und Jagdökonomie der Ahrensburger Rentierjäger im Mittelgebirge. Röm-Germ Zentralmus Monogr 1996; 38: 1-364.

[29] Manker E. Lapparnas heliga ställen. Acta Lapponica 1957; 13: 1-462.

[30] Sarmela M. Finnische Volksüberlieferung. Atlas der finnischen Volkskultur 2. Münster-New York-München-Berlin, Waxmann-Verlag 2000; p. 410.

[31] Mebius H. "Värro", Studier i Samernas förkristna Offerriter. Skrift utgiv relig inst Uppsal Hum fak Genom, 1968: p. 5.

[32] Ehrhardt KJ. Alte Kultsleine und Opferplätze der finnischen Lappen im Gebiet des Inarisees und lijärvi. Anthropos 1964; 59: 843-8.

[33] Diedrich C. Ice Age geomorphological Ahorn Valley and Ailsbach River terrace evolution– and its importance for the cave use possibilities by cave bears, top predators (hyenas, wolves and lions) and humans (Late Magdalénians) in the Franconia Karst – case studies in the Sophie's Cave near Kirchahorn, Bavaria. Quat Sci J 2013; 62(2): 162-74.

[34] Bosinski G. Les figurations féminines de la fin des temps glaciaires. In: Musée National Préhistoire Les Eyzies de Tayac (Ed.). Mille et une femme(s) de la fin de temps glaciaires 2010; pp. 50-67.

[35] Gumpert K. Der Madeleinzeitliche Rennerfels in der Fränkischen Schweiz. Praehist Z 1981; 22: 1-77.

[36] Sommer CS. (Ed.). Archäologie in Bayern - Fenster zur Vergangenheit - zusammengestellt von 25 Jahre nach Gründung der Gesellschaft für Archäologie in Bayern. Verlag Friedrich Pustet 2006; p. 336.

CHAPTER 8

MAIN POST-LGM SPELEOTHEM PERIOD

Abstract: Within the climatic change after the LGM to the final Ice Age interstadials and stadials changed with humid and dryer periods. A strong "cave ceiling collapse" throughout the cave within this time frame blocked some passages up to the Collapse Hall. Most of the candle stalagmites were formed in this time between 16.000-12.000 BP with the last and main speleothem genesis, which continues since the Holocene. The cave floor was covered especially in the larger halls by falling large blocks which scattered the reindeer antlers in the Reindeer Hall, but also closed the connections of the chambers/halls in the middle part of the cave. Between those, and on those, a variety of different speleothem types formed, whereas the largest are found within the Millionary Hall. Sinter basins are also found there and in the connected Reinder Hall, only. After all sedimentological, stratigraphic, and cave bear clock dating methods, nearly the complete large "sinter decoration" of the Sophie's Cave must have build up in the middle to late Late Pleistocene covering the cave bear skulls/bones of *U. spelaeus eremus/spelaeus* and *U. ingressus* and their bonebeds all over in a second larger speleogenesis time starting in a warmer interstadial around 42.000 BP (Millionary after Chapter 3) and a third final main speleogenesis time around 16.000-12.000 BP in the Alleroed to Dryas which continued since the Holocene.

Keywords: Speleothem forms, ages, climate record, cave part and speleothem layer collapses, final Late Pleistocene, Post-LGM.

CAVE COLLAPSE/SPELEOTHEMS - END OF THE LATE PLEISTOCENE (POST-LGM, MIS 1)

The few larger speleothems such as the Elefantenohr (= Elephant Ear), Bienenkorb (= Bee Basket), Kleiner und Großer Millionär (= Small and Large Millionary, Fig. **1**), grew already before the latest speleogenesis time and are much younger, as always told by cave guides: "one million years old". Those can be estimated at the moment without absolute dating only not be older than 42.000 BP (Bölling Interstadial), because they also cover the small cave bear species *U. spelaeus eremus/spelaeus* bonebed in the Reindeer and Millionary halls, which cave bear subspecies that date themselves with the "cave bear clock" into the early/middle Late Pleistocene (see dating by teeth: Chapter 3, Fig. **2**) [1]. How-ever, those must have continued to grow also in the Post-LGM during the latest main speleogenesis times and continue to grow until today.

With the end of the Late Pleistocene, some thousands of years after the peak of the LGM around 19.000 BP [2], the climatic change of the warmer and colder interstadials of the Dryas to Alleröd periods caused an increase in new ground-water in Central Europe [3]. The temperature and humidity changes were res-ponsible for a new speleothem genesis period in the Sophie's Cave, such as also documented and dated in the nearby Zoolithen Cave [4, 5].

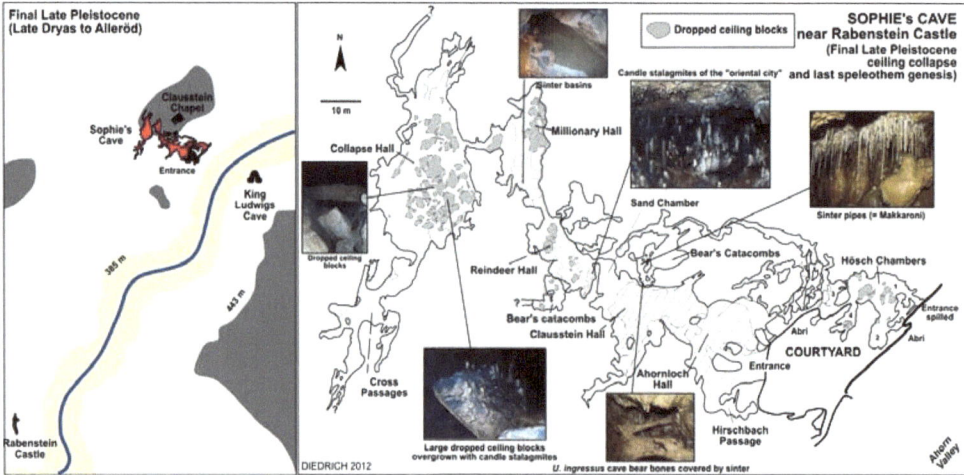

Figure 1. Transition final Late Pleistocene to Holocene (dry, in the Dryas to Alleröd periods, app. 16.000-14.000 BP) – main cave ceiling collapse in the third and main speleothem genesis period.

The formerly dry Sophie's Cave became wet in many parts, causing a massive ceiling collapse (Fig. **1**). The larger the halls, the more large blocks dropped from the ceilings. This is best visible on the largest block dump in the centre of the Collapse Hall (= grey coloured mapped blocks in Fig. **1**). This obviously had no effect on humans or larger animals anymore. The blocks themselves, and other places and even the cave bear bonebeds were completely covered by speleothems, especially the 5-10 cm in diameter thin candle stalagmites (Figs. **2**, **3C**), which dating is herein only correlated to other absolute dated candle-like and similar in diameter sized speleothems of Upper Franconia caves, especially to those of the

Zoolithen Cave [4]. Speleothem dates (Uranium/ Thorium) of a candle stalagmite sample of the Zoolithen Cave bone breccias gave ages on the upper and last speleothem layer of about 11.720±125 BP (older data from 1950 [6]) which is calibrated 13.720±125 BP. These seem to have built not only in the Zoolithen Cave during the latest Upper Pleistocene Alleröd climatic change to a warmer period (13.500 - 12.700 BP [4]). Also in other Upper Franconia caves those are found with absolute dating, such as the Sophie's Cave or Große Teufels Cave. The cover by the youngest speleothems all over the Sophie's Cave "floor" of the ceiling blocks, and bonebeds, some partly articulated small cave bear *U. s. eremus* skeletons (in the Reindeer and Millionary halls) or even the absolute dated reindeer antlers/mammoth coxa correlate to the dating of those as Pre-LGM deposited megafauna remains. All three main speleothem generations can be explained only with the general cave model, that includes four main cave genesis phases based on the stratigraphy, sedimentology and dated cave bears (Fig. **2**).

Sinter Tubes (= Makkaroni)

Speleothems developed from the leachate, originating from rain waters which absorbed the CO_2 [7]. Within the soil they built an acid that dissolved the carbonates ($CaCO_3$), including dolomites (= $CaMgCO_3$), which the waters transport into the underground and its cavities [7, 8]. The water ran within tiny to larger clefts of the Sophie's Cave Jurassic reef and carbonate sand dolomites (Fig. **2**). During its appearance within the cavities, speleothems developed because CO_2 disappeared quickly due to pressure and temperature differences within the caves [7, 8]. The diffusion of the gas from the water drops caused a remineralization of the carbonate, being the initialization of the speleothem genesis [9 - 12]. Those crystals are colored by red Iron and black Manganese ions, but in some cases, white speleothems consist of 100% pure carbonate [11, 12]. The best well-known variations in speleothems and their types well known from caves all over the world [7 - 12] can be demonstrated also with the main large types in the Reindeer and Millionary halls best (Figs. **1**-**7**).

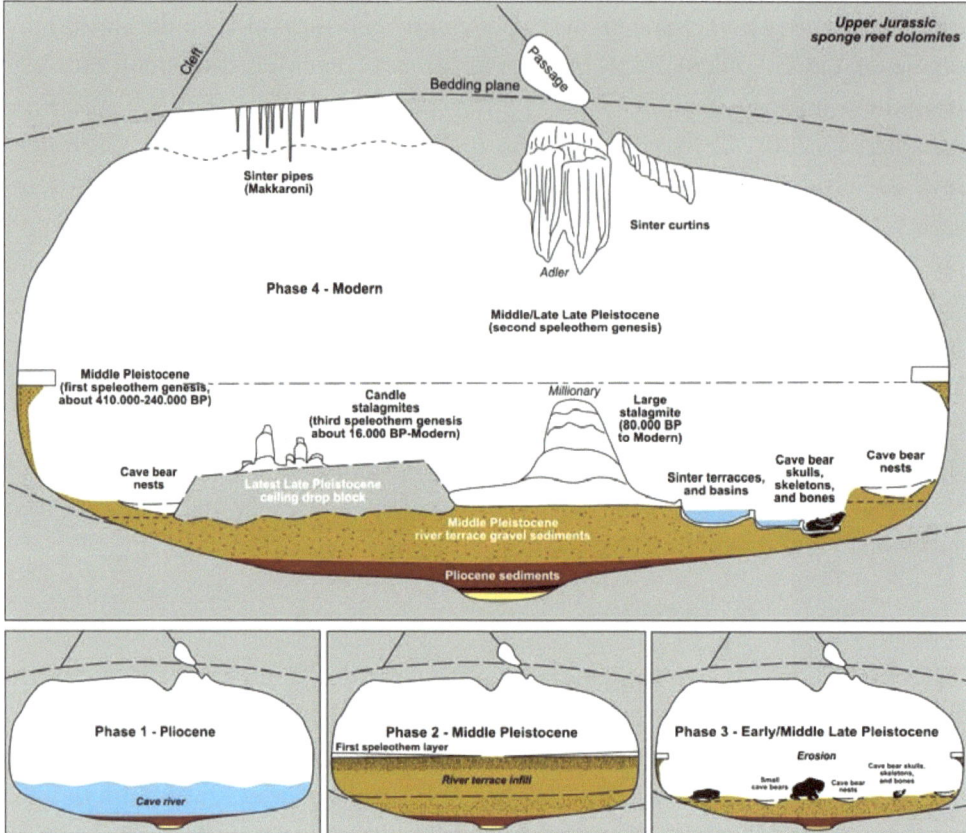

Figure 2. Schematic cross section through the Millionary Hall and Reindeer Hall in which typical speleothem types are represented (cf. photos of Figs. 3-7). From these, four cave genesis phases can be reconstructed including three main speleothem periods. Phase 1- Cave river in the Pliocene and first "colored series" sediments. Phase 2. Infill of first Ante-Ailsbach River terrace sediments which left massive "yellow series" and coarse gravels and first speleothem layer. Phase 3 – Further erosion of the Middle Pleistocene terrace sediments and speleothem layer, and after use of the hall by small cave bears for hibernation, within the early/middle Late Pleistocene and second speleothem genesis (around 42.000 BP). Phase 4 - Recent stage after latest late Pleistocene further ceiling collapse, and final last speleothem layers and build up of mainly candles and sinter terraces.

First, water dripping from the ceiling creates a ring-structure of carbonate. From this only 5 mm wide and mainly between 1-30 cm long pipes build up, which have typical growth rings [11] (Fig. **3A**). Within this Makkaroni the flowing water is responsible for further rings and the pipe growth [11] (Fig. **3A**). Such pipes only grow in lengths, and often crack due to water loss within the rocks, their own weight, by cave bears or other animals, and even by cavers and earth quakes [11]. These Makkaronies are well preserved in the lower part of the Reindeer Hall (Fig. **3B**) and are found commonly as fragments in the cave sediments of the upper series starting in the cave bear bonebeds.

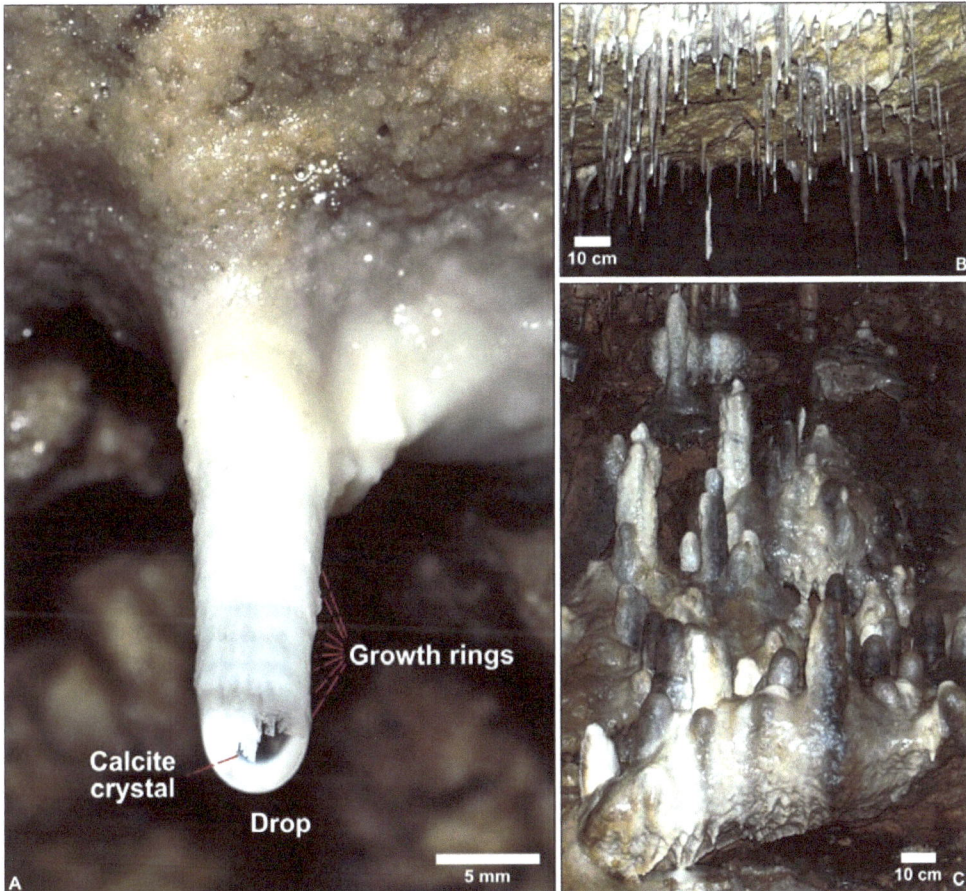

Figure 3. A-B. Sinter tubes (= Makkaroni) on the ceiling, its growth, and **C.** Candle stalagmites as antagonists on the floor of the upper part of the Reindeer Hall, Sophie's Cave.

Stalagmites

These are the antagonists of the stalactites or Makkaroni, and grow opposite, from the ground [11] (Fig. **4C**). In general, opposite the stalagmite is a Makkaroni, or stalactite on the ceiling, such as with the best example in the Sophie's cave with

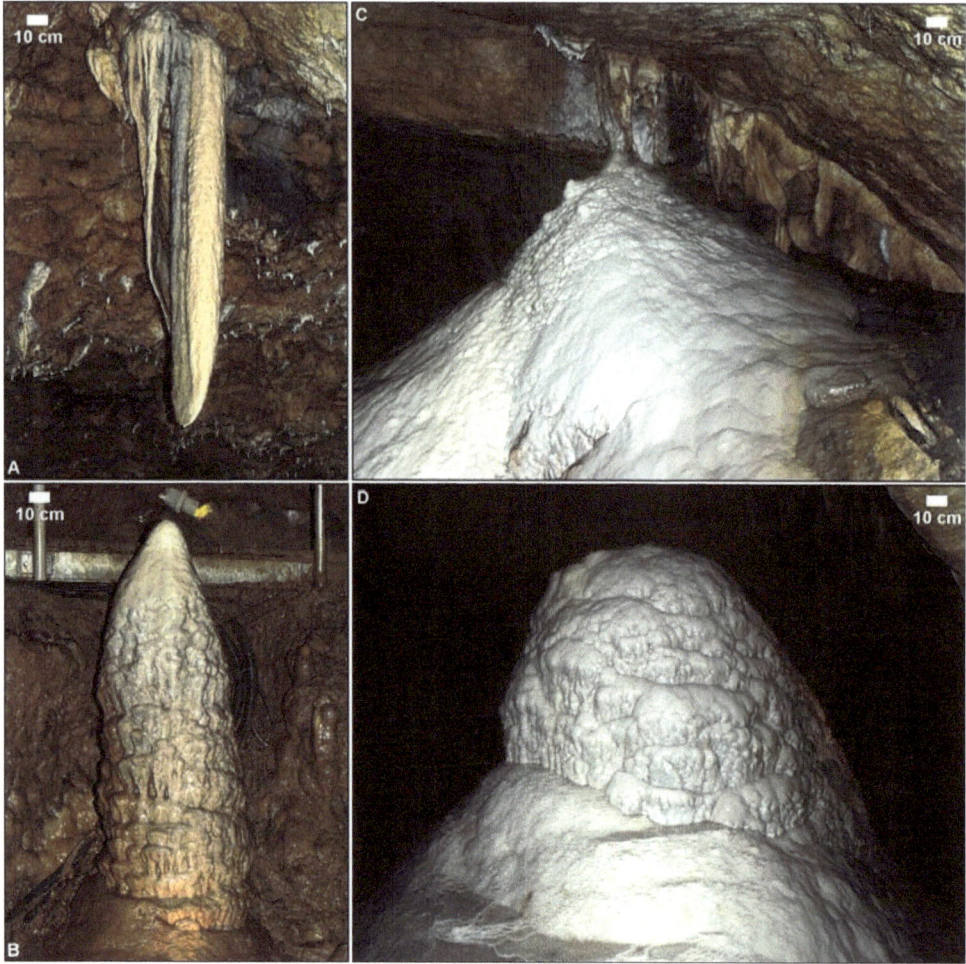

Figure 4. Stalactites and stalagmites in the Reindeer and Millionary Halls. **A-B.** "Elefantenohr" and Stepp-like antagonist "Bienenkorb" (= Bee Basket) in the Reindeer Hall. **C.** Small Milllionary stalagmite and **D.** Large Millionary stalagmite in the Millionary Hall, Sophie's Cave.

the "Elephantenohr and Bienenkorb" (Fig. **4B**). Their growth depends on the amount of dripping waters. The water drips onto its tip, splashes into very tiny drops which then lose some CO_2 directly crystalizing the carbonate which give the typical cascade-like form [10]. Such variations of candle-like stalagmites are all over most of the Sophie's Cave, with larger amounts of more regular straight and 5-10 cm in diameter thin ones in the upper Reindeer Hall, especially densely in the "Orientalischen Stadt" (Fig. **5**).

From the few drops, the speleothems grow, especially in height [11], and can range up to several meters (*e.g.* observed in the Zoolithen Cave, Große Teufels Cave). These seem to have built up in the Sophie's Cave and other caves within the past 16.000 years, as discussed above and in the climatic and general valley /cave development models of the cave [1]. With permanent, more massive drippings, or flowing water, wider forms develop, much more quickly [12].

Best examples are the small (Fig. **4C**) and large Millionary (Fig. **4D**) in the Millionary Hall, which seem to be not older as 80.000 BP at least in their main speleogenesis (first Late Pleistocene interstadial), continued in growth about 42.000 BP (second Late Pleistocene interstadial) with further growth during the Post-LGM. Their Late Pleistocene age is also explained in Fig. **2**, because those directly cover the early/middle Late Pleistocene cave bear bonebeds (small *U. s. eremus/spelaeus* forms). Its thickness was also estimated with its modern growth. Every decade, a one to two mm thin sinter layer is built up on the Large Millionary, which was proven by a "cemented" old visitor ticket. Using this slow growth in humid/warm periods also estimated at other cave sites in Europe [10 - 12] and calculating this with time to be about 25 cm in maximum is expected to have grown mainly since the Post-LGM, which includes permanent sipping water.

The climate within the second main speleogenesis period was even more humid, and much more water must have flowed over this large stalagmite, which could have allowed it to grow much faster.

Stalactites

From the Makkaroni, sometimes stalactites form on the ceilings, when the pipe is blocked in the canal by dirt or carbonate [11].

In such cases, the water then runs outside the pipe and the Makkaroni grows from outside [11]. From such stalactites even irregular forms can develop as sinter curtains [11, 12], such as the "Elephantenohr" in the Reindeer Hall (Fig. **4A**) or the "Adler" (= Eagle) in the Millionary Hall (Fig. **6A**).

Stalagnats

If a stalagmite and stalactite fuse, such a form develops [11, 12]. There are a few such known in the Sophie's Cave with the best example nearby the Elephantenohr

Figure 5. A. Stalagnat in the Reindeer Hall, Sophie's Cave.

in the Reindeer Hall (Fig. **5**). Even the Elephantenohr and its antagonist Bienenkorb (Fig. **4A-B**) are already close to each other and will most probably build up in the future a stalagnat.

Sinter Curtains

Their genesis is different, not resulting from Makkaronies but from water running on angeled cave wall surfaces, thereby producing an elongated carbonate trail [11, 12]. This normally is sinus-like, developing a different carbonate crystal type without parallel surfaces, which both influence the curtain speleothem shape [11, 12]. Those forms are rarer in the cave and present with largest and best preserved specimens in the Reindeer and Millionary Halls (Fig. **6**).

Figure 6. A. Sinter banners with the "Adler" (= Eagle) and **B-C.** Sinter and curtains in the Millionary Hall, Sophie's Cave.

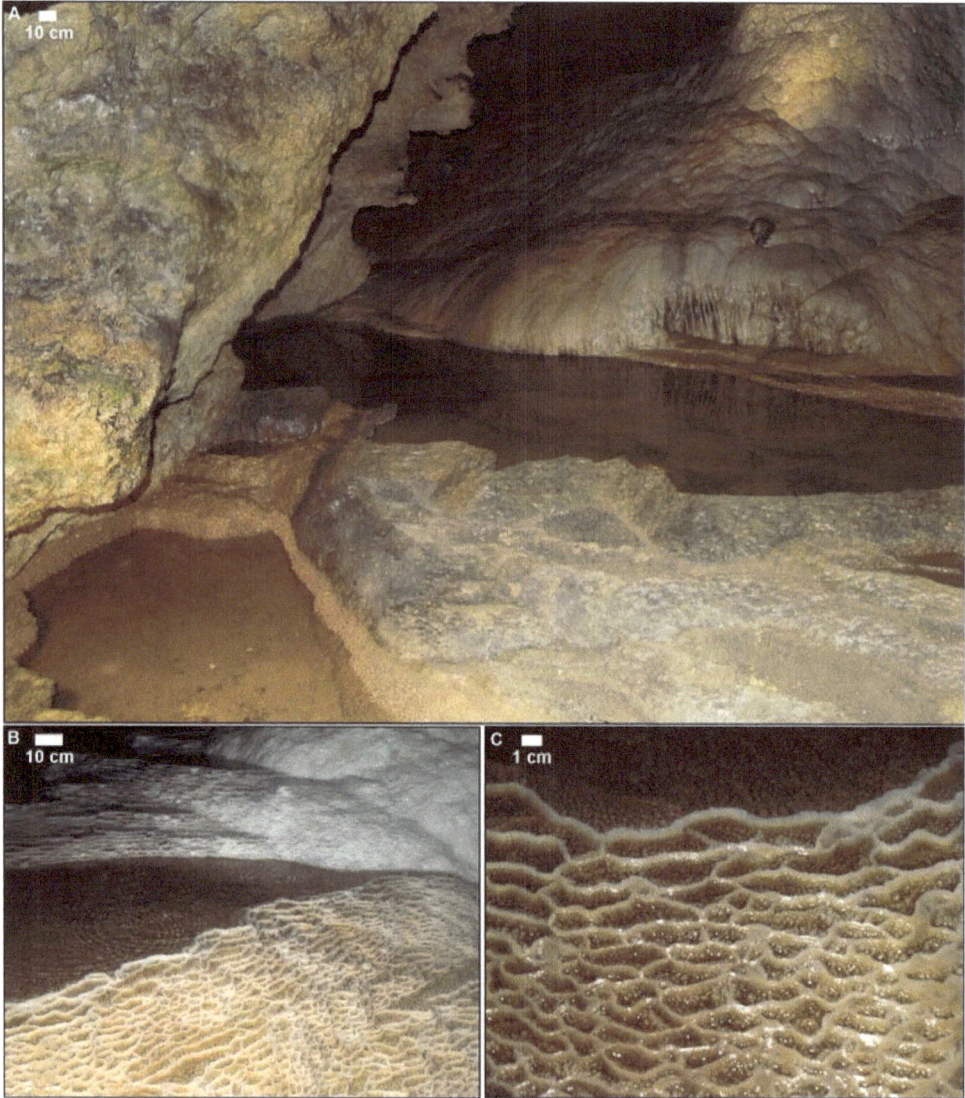

Figure 7. A-B. Sinter terraces and basins, and **B-C.** Microbasins starting from the Large Millionary stalagmite in the Millionary Hall, Sophie's Cave

Sinter Basins

These are planed forms on the floor only [11, 12]. Non-active dry ones which are much older and of the first speleothem genesis are still present in the upper hidden part of the Clausstein Hall, but the main and active ones from the younger speleothem period are in the Millionary Hall (Fig. **7A**), which were mapped in more detail (see Chapter 4, Fig. **12**). There, the terraces are still active, and depending on seasonal waters some basins are not filled. However, during a cleaning process in 2011, several damages (due to stepping onto those since 1833) were found within basin steps, which were filled somehow with cave mud, to keep water in the each of the basins, also to protect or support the habitat of small water organisms (see Chapter 5). The largest basins are near to the Large Millionary (cf. map in Chapter 3, Fig. **7A**). Small microbasins of one to several centimeters in size only, are close to the large Millionary (Fig. **7B-C**).

REFERENCES

[1] Diedrich C. Ice Age geomorphological Ahorn Valley and Ailsbach River terrace evolution– and its importance for the cave use possibilities by cave bears, top predators (hyenas, wolves and lions) and humans (Late Magdalénians) in the Franconia Karst – case studies in the Sophie's Cave near Kirchahorn, Bavaria. Quat Sci J 2013; 62(2): 162-74.

[2] Clark PU, Dyke A, Shakun JD, *et al.* The Last Glacial Maximum. Science 2009; 325(5941): 710-4.

[3] Richter DK, Götte T, Niggemann S, Wurth G. REE3+ and Mn2+ activated cathodoluminescence in late glacial and Holocene stalagmites of central Europe: evidence for climatic processes? The Holocene 2004; 14: 759-67.

[4] Kempe S, Rosendahl W, Wiegand B, Eisenhauer A. New speleothem dates from caves in Germany and their importance for the Middle and Upper Pleistocene climate reconstruction. Acta Geol Polon 2002; 52(1): 55-61.

[5] Diedrich C. Holotype skulls, stratigraphy, bone taphonomy and excavation history in the Zoolithen Cave and new theory about Esper's "great deluge". Quat Sci J 2014; 63(1): 78-98.

[6] Poll KG. Die Zoolithenhöhle bei Burggaillenreuth und ihre Beziehung zum fränkischen Höhlen- und Kluftsystem. Erl Forsch B Naturwiss 1972; 5: 63-76.

[7] Bretz JH. Vadose and phreatic features of limestone caverns. J Geol 1942; 50: 675-811.

[8] Ford DC, Williams PW. Karst geomorphology and hydrology. London: Unwin-Hyman 1989; p. 601.

[9] Klimchouk AB. Hypogene Speleogenesis: Hydrogeological and Morphogenetic Perspective. Carlsbad NM Spec Pap 2007; 1: 1-106.

[10] Jennings JN. Karst Geomorphology. Oxford: Blackwell 1985; p. 293.

[11] Self CA, Hill CA. How speleothems grow: An introduction to the ontogeny of cave minerals. J Cave Karst Stud 2003; 65(2): 130-51.

[12] Fairchild IJ, Frisia S, Borsato A, Tooth AF. Speleothems. In: Nash DJ, McLaren SJ. (Eds.) Geochemical Sediments and Landscapes. Oxford: Blackwell 2006; pp. 200-45.

CHAPTER 9

POSTGLACIAL ARCHAEOLOGY

Abstract: Postglacially, humans used the entrance throughout the Urnenfeld Bronze Age into the Hallstadt/La Tène Iron Age, which is documented by pottery shards. The lower cave entrances were occupied intensively at the Early Medival times during which the castle was built 30 meters above the cave entrance on the Jurassic dolimite plateau. From this time period, few animal and human bones such as iron/bone/stone tools and bone kitchen rubbish and several pottery shards were found on the yard infront of the today's cave entrance and the lower cave part entrance areas.

Keywords: Few archaeological finds, Urnenfeld Bronze Age, Hallstadt/La Tène Iron Age, pottery, Early Medival castle, cave entrance use.

BRONZE AGE (AROUND 3.350 BP)

Bronze Age tools and ceramic fragments from the Early Bronze Age (= Hügelgräberzeit) were already mentioned in historical reports within the area of the Clausstein Hall [1, 2]. The oldest archaeological remains date into the Late Bronze Age (= Urnenfelderzeit, Hallstadt [3 - 5]). Surface finds were collected and rescued only from the Hösch Chamber 4 (Fig. 1). From this, about 30 quartz sand tempered typical prehistoric pottery ceramic fragments (with inner side black and outer side red) were saved against future trampling. Most of the material was not taken and left "in place". Material was already "deposited" before by modern speleologists in niches to protect them against trampling destruction. A single marginal shard fragment with finger print decoration was found sticking between blocks at the end of the diagonal shaft (Figs. 1, 2A). This decorated shard date into the Late Bronze Age (about 3.350 BP) [1, 6]. It is a fragment from a large storage vessel, which is almost 50 cm in height [1]. In such, cereals, peas, lentils or beans were stored [1].

Another shard with slicked mud decoration and with sinter skin and rhomboidal decoration (= Dellenstrich [1]) dates into this time, too. All pottery types date more into the Late Bronze Age, but partly existed longer into the Iron Age [6]. When the pottery was only "daily use ceramics" possibly it was thrown into Chamber 4, from the courtyard, into the small diagonal shaft for ? sacrifices, or this part was a "storage site" to keep food fresh.

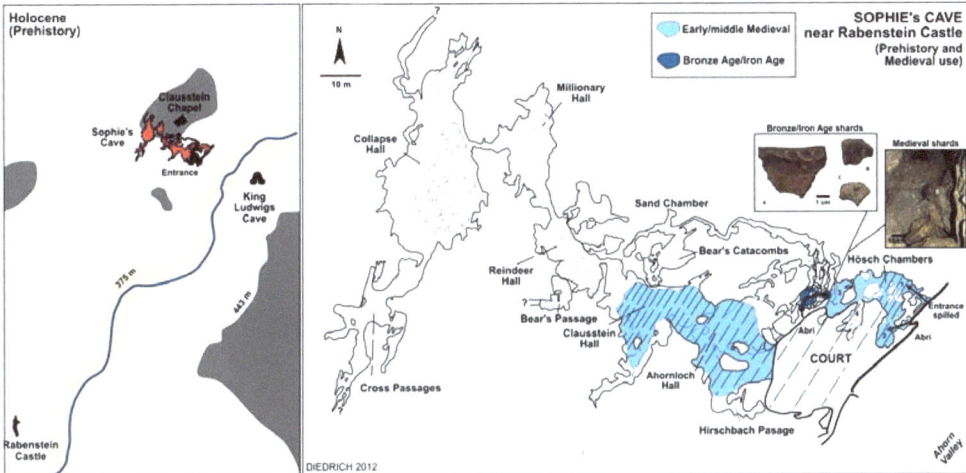

Figure 1. Prehistoric use and Early Medieval Age use of the Sophie's Cave areas.

HALLSTATT/LA TÈNE IRON AGE (AROUND 2.600 BP)

Some coarse shards from the Hösch chambers 1 and 2, and possibly also from the Ahorn/Clausstein Halls, are left by people of the Early Iron Age (Hallstadt/La Tène Cultures [3 - 5] (Figs. **1**, **2**). Additionally, in 2007, shards from the archaeological trenches in the court area in front of the entrance seem to indicate the use of the anterior cave parts in prehistoric times. Secondary burials in vertical shafts are known from the nearby Esper Cave and Zoolithen Cave [7] near Burggailenreuth, although this is not proof in the Sophie's Cave. The absence of human bones or burial pottery in the Sophie's Cave supports the theory of a non-burial place of unclear use by Iron Age people.

THE CAVE USE DURING MEDIEVAL TIMES (11-13 CENTURY)

In the courtyard of the Sophie's Cave entrance area, quite abundant remains of pottery shards, bones, and even metal tools were collected, which are typical for Early-Late Medieval castles [8]. It is unclear, if those were dropped from the fortifications above, or are from cave use purposes. In contrast to the modern influenced archaeological horizons/cave areas, these seem to be more or less undisturbed within the Hösch Chambers 1-2. There, martens, foxes and badgers caused bioturbation and the possibility of surface finds (see Chapter 10, Fig. **1**). In the Late Medieval times (11-13 Century), the time when the Ahorn Castle existed 40 meters above the modern cave entrance [9], the Hösch Chambers and Ahornloch Hall must have been used for living, storage, animal shelter or during war times for protection of people. Metal remains are billets, door jacks, nails or knife fragments (Fig. **3A-D**). The ceramics consist of the abundant typical

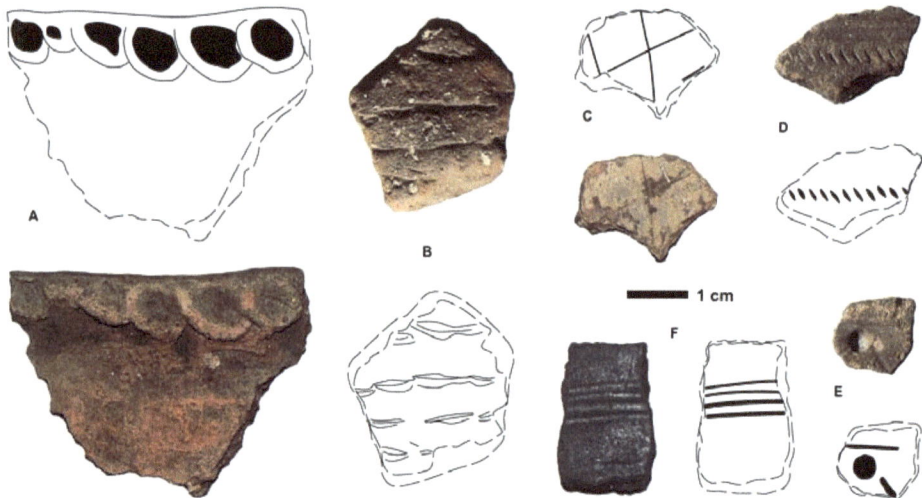

Figure 2. Decorated ceramic fragments from the prehistory (Bronze to Iron Ages) from the "Sophie's Cave" and mainly its courtyard. **A, D, E.** Shards with finger print decorations (Urnenfeld Bronze Age, app. 3.350 BP). **B.** Shard (La Tène Age B). **C.** Shard with feldspar temper, sinter skin and grid decoration (La Tène Age D). **F.** Shard with parallel ornamentation Hallstadt Age D3) (coll. Rabenstein Castle Museum).

"kitchen pottery", which was burned in ovens to the typical "medieval grey ware" [10 - 14]. This non-decorated and non-colored pottery types, which were tempered with fine quartz sand, are mostly cooking pots and bowls [10]. Those are represented in the cave entrance yard surface area collected materials with some marginal shard fragments (Fig. **3**). In Chamber 4, the remains of a smashed grey and burned pottery (probably lamp function) was found and left *in situ* (Fig. **1**). Other finds are a fine-sand grinder (Fig. **3E**), and typical Early Medieval wheel symbol decorated horn comb fragment [14, 15] (Fig.

3F). The few colored or relief-like decorated harder burned ceramic fragments (Fig. **3I-J**) date into the Late Medieval times, as compared *e.g.* with other German castle sites [10]. The Medieval domestic animal remains consist of teeth and smashed/cutted bones of a small cattle type, pigs, and sheeps or goats [16, 17] (Fig. **3K-N**). Cut and smashed bone remains of farm animals (cattle, pig, sheep/goat) were also excavated within the courtyard trench in 2007 [18]. Finally, there is a femur from a large, modern horse (most probably not Medieval times), which was found somewhere in the anterior cave area.

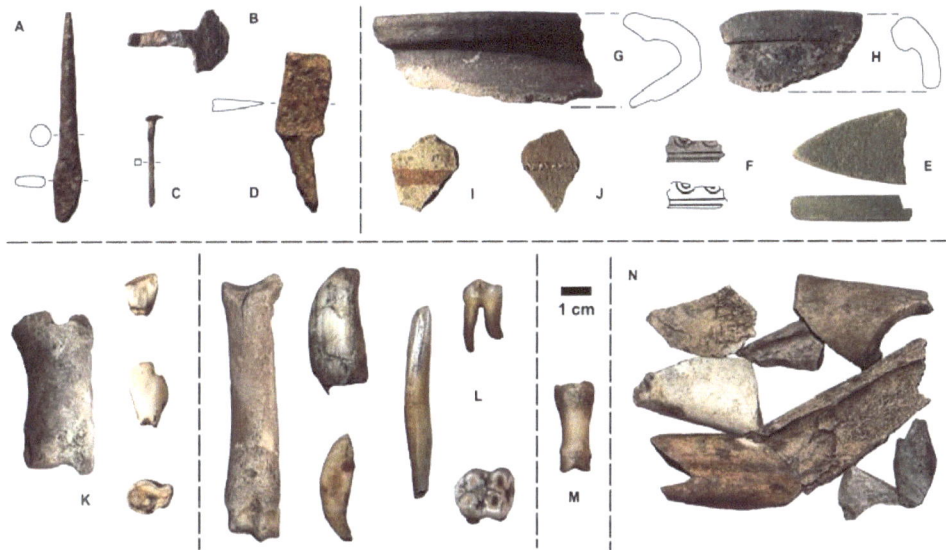

Figure 3. Finds from the early to middle Medieval times (time of Ahorn Castle above the cave). **A.** Billet, **B.** Door jack, **C.** Nale, **D.** Knife fragment, **E.** Fine-sand grinder, **F.** Horn comb decorated tarnish with wheel symbols, **G.** Globular pot marginal shard (grey ware), **H.** Pot marginal shard (grey ware), **I.** Shard (Late Medieval), **J.** Shard (Late Medieval), **K.** *Bos* (small Medieval cattle): Phalanx I, Incisor and molar teeth of calves. **L.** *Sus* (house pig) metapod, teeth of the upper and lower jaws, **M.** *Ovis/Capra* (sheep/goat): Phalanx I, **N.** Smashed long bones (fragments) (coll. Rabenstein Castle Museum).

REFERENCES

[1] Jockenhövel A, Kubach K. (Eds.). Bronzezeit in Deutschland. Hamburg, Nikol Verlagsgesellschaft 2000; p. 111.

[2] Staatsarchiv Bayreuth, unpublished documents K3FVVIII/321. Regierung des Obermainkreises/Oberfranken 1837-44.

[3] Werner T. Hallstattkultur. Göttinger Typent Ur- Frühgesch Mitteleur 1984; pp. 1-57.

[4] Abels B-U. Die vorchristlichen Metallzeiten. In: Abels B-U, Sage W, Züchner C. (Eds.) Oberfranken in vor- und frühgeschichtlicher Zeit, 2 Auflage. Bamberg, Bayerische Verlagsanstalt 1996; pp. 65-160.

[5] Stöllner T. Kulturwandel, Chronologie, Methoden. Ein Diskussionsbeitrag am Beispiel der Hallstatt- und Latènekultur. Prähist Z 1999; 74: 194-218.

[6] Schreg R. Keramik aus Südwestdeutschland. Eine Hilfe zur Beschreibung, Bestimmung und Datierung archäologischer Funde vom Neolithikum bis zur Neuzeit. Fundber Baden-Württemberg 1999; 23: 385-617.

[7] Diedrich C. Holotype skulls, stratigraphy, bone taphonomy and excavation history in the Zoolithen Cave and new theory about Esper's "great deluge". Quat Sci J 2014; 63(1): 78-98.

[8] Sage W. Frühgeschichte und Frühmittelalter. In: Abels B-U, Sage W, Züchner C. (Eds.) Oberfranken in vor- und frühgeschichtlicher Zeit, 2 Auflage. Bamberg, Bayerische Verlagsanstalt 1996; pp. 161-280.

[9] Schwarz K. Die vor- und frühgeschichtlichen Geländedenkmäler Oberfrankens. Materialh Bayer Vorgesch 1955; 5(1): 1-203.

[10] Losert H. Die früh- bis hochmittelalterliche Keramik in Oberfranken. Band 1: Text, Katalog, Band 2: Tafeln. Z Archäol Mittelalt Beih 1998; 8: 1-221.

[11] Lüdke H, Schietzel K. (Eds.) Handbuch zur mittelalterlichen Keramik in Nordeuropa. Ethnogr-Archäol Z 2002; 43: 487-94.

[12] Bauer W. Zur Herstellung mittelalterlicher Kugeltöpfe. Z Ver hess Gesch Landesk, 1954-55; 65/66: 243-7.

[13] Gross U. Frühmittelalterliche Siedlungskeramik aus dem Taubertal. Ber Röm-Germ Kommis 2006; 87: 1-76.

[14] Falk A. Knochengeräte des späten Mittelalters und der Neuzeit. Z Ver Lübeckische Gesch Altertumsk 1983; 63: 105-28.

[15] Richter U. Mittelalterliche Knochenkämme aus Freiberg. Ausgr Funde 1990; 35: 37-40.

[16] Hüster H. Untersuchungen an Skelettresten von Rindern, Schafen, Ziegen und Schweinen aus dem mittelalterlichen Schleswig (Ausgrabung Schild 1971-1975). Ausgr Schleswig Ber Stud 1990; 8: 1-137.

[17] Hüster H. Einflüsse ökologischer Faktoren auf die Körpergröße von Wiederkäuern im Verlauf des Mittelalters. Dt Ges Säugetierk 1991; p. 20.

[18] Steguweit L. Untersuchungen der Schichtenfolge am Vorplatz der Sophie's Cave bei Burg Rabenstein. In: Greipl EJ, Sommer S, Päffgen B. (Eds). Das archäologische Jahr in Bayern 2008. Theiss-Verlag 2008; pp. 11-3.

MODERN CAVE ANIMALS AND GUESTS

Abstract: Today, some possibly endemic cave animals inhabit or use seasonally different parts of the cave system, which is still in use as a common fox den reaching hereby deeper cave parts of the lower cave system. Common cave spiders (*Meta menardi*) or moths (*Triphosa dubitata*) such as rare bats (*Myotis myotis*) use only the larger first two halls of the cave entrance area, whereas bats are rarely found for hibernation only deeper in the cave. In the middle part of the cave, water bodies of the speleothem terrace basins contain possibly endemic small crustaceans (*Bath-ynella*), in the mud infaunistic pigment and eye-less flatworms (*Phagocatta*) or on the water surfaces springtails (*Heteromurus/Onychiurus*), which species are not yet determined.

Keywords: Modern vertebrates and invertebrates, fox den, bat hibernation, spider and moths in anterior cave entrance halls, possibly endemic fauna, crustaceans and eyless flatworms in speleothem water basins.

COMMON/RED FOX DEN

The Hösch Chambers of the lower cave area were used in modern times as a common (or red) fox den (Fig. **1**). Cave use is typical for *Vulpes vulpes* all over Europe [1]. Remarkably, this fox den was in use only for the past three years, reaching deep into the Hösch Chambers, and finally into the Bear's Catacombs. In Chamber 2, a fox cloak was discovered, which fresh excrements indicates use for the past years. Also several bones of domestic animals (cattle, sheep, and goat), and wild animals (roe, boar, hare) (?plus remains of a raccoon) have been imported by the medium-sized carnivore very deep and into the Bear's Catacombs, which denning behaviour is typical in fox dens [1, 2]. The modern, light-yellowish colored prey bones were accumulated only in some areas. Also the faecal areas are massive only at two points (Fig. **1**). Astonishing are the completely scratched walls along the passage between the Ahornloch Hall vertical shaft and Bear's Catacombs (Fig. **1**). Obviously a fox (?or the raccoon) was caught for a time in the Bear's Catacombs. By trying to escape it left in places 2 meter high on the walls each 3-4 parallel and short claw scratch marks. At another place the carnivore jumped high enough (also scratch marks) through a diagonal shaft to reach the Hösch Chambers 3 and 4.

BATS, SPIDERS, MOTHS, WORMS AND CRUSTACEANS

The Ahornloch/Clausstein Halls and Hösch Chambers contain most of the modern cave animals, especially bats, insects or snails (Fig. **1**) with about 35 different invertebrate species records [3]. This cave is one of the richest Upper Franconia cave in modern biodiversity [3].

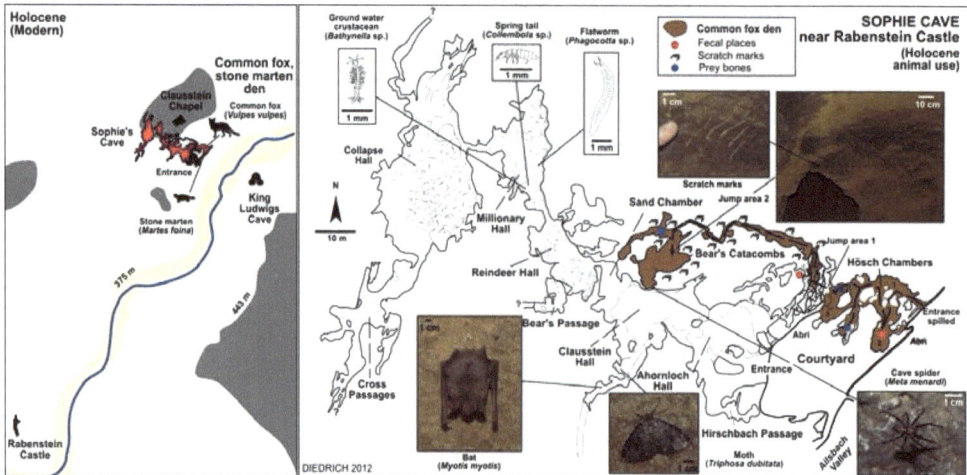

Figure 1. Modern bats, spiders, moths, worms, crustaceans, and common/red fox (*Vulpes vulpes*). At important crossings, foxes left faecal places for orientation. Its prey bones were enriched in some places. Interesting is the abundance of scratch marks (four parallel) on the dolomite wall of one passage to and in the Bear's Catacombs Chamber itself. At two places the foxes must have jumped up to two meters high to find an "exit" (Hösch Chambers 3 and 4 and to Bear Catacombs).

The complete Sophie's Cave is not a typical bat hibernation cave at all [3]). It is used occasionally, but only in the anterior areas, such as documented in 2011 for two specimens of *Myotis myotis*, the common European water bat [4] (Fig. **2A-B**).

In the anterior cave areas are mainly seasonal animals (= subtroglophils [3]. To this group belongs the very common cave moth *Triphosa dubitata* (Fig. **2C**) that hibernates in large amounts, starting in the late summer [3]. Another common insect is the cave spider *Meta menardi* [3] (Fig. **2D**) which hides in cavities in the

Figure 2. Most common modern cave animals in the Clausstein Hall, Sophie's Cave. **A-B.** One and two *Myotis myotis* before their hibernation. **C.** *Triphosa dubitata*, dark variant. **D.** *Meta menardi*, smaller male. **E.** *Meta menardi* cocoon.

walls and ceilings. On those walls are often attached some centimeter long, drop-like cocoons (Fig. **2E**). From those more than a hundred juveniles hatch, of which some leave the cave to find new habitats [3].

Also common are about one centimeter large larvae of the cave mushroom bug *Speolepta leptogaster* which as juveniles have poor eyesight [3]. The larvae live on a fine web which has slimy fine drops [3].

In the deeper non-photic cave areas, such as the Millionary Hall, only very small animals live there permanently (= eutroglophils [5]) within or on the waters, or in the mud within the sinter basins. On the water surfaces jumping white springtails *Heteromurus* and *Onychiurus*can can be seen well [3, 6]. Although *Onychi-urushas* has reduced spring tails, *Heteromurus* really can jump like a flea [6]. Deep in the cave are eutroglobionts [5]: crustaceans and worms adapted to permanent dark-living. The one milimeter small crustacean is called *Bathynella* sp. which species has not been determined yet [7]. The worms are represented by a pigment and eye-less flatworm of the genus *Phagocatta* [8] (Fig. **1**). This seems not to be "old" and "endemic" as suggested by Schabdach [3], because they live in the modern (since 1833 produced along the trails) mud of the sinter basins. Those might have been probably imported by eggs sticking in the mud of the shoes of cave visitors, especially spelunkers. Very obviously the appearance of this infauna is limited to two water filled and modern influenced sinter basins. Only in the upper soft mud and not in the lowermost and "black-organic-rich" layers such worms are found in colonies.

REFERENCES

[1] Labhardt F. Der Rotfuchs. Naturgeschichte, Ökologie und Verhalten dieses erstaunlichen Jagdwildes. Hamburg: Paul Parey-Verlag 1990; p. 158.

[2] Görner M, Hackethal H. Säugetiere Europas. Stuttgart: Enke-Verlag 1988; p. 371.

[3] Schabdach H. Die Sophie's Cave im Ailsbachtal. Wunderwelt unter Tage. Ebermannstadt: Verlag Reinhold Lippert 1998; p. 47.

[4] Schober W, Grimmberger E. (Eds.) Die Fledermäuse Europas. 2 Auflage. Stuttgart, Kosmos 1998; p. 265.

[5] Weber D. Die Evertebratenfauna der Höhlen und künstlichen Hohlräume des Katastergebietes Westfalen einschließlich der Quellen- und Grundwasserfauna. Abh Karst-Höhlenk 1991; 25: 1-701.

[6] Hopkin SP. Biology of the Springtails (Insecta: Collembola). New York 1997; p. 330.

[7] Gledhill T, Sutcliffe DW, Williams WD. British Freshwater Crustacea Malacostraca: A key with ecological notes. Freshw Biol Ass Sci Publ 1993; 52: 1-173.

[8] Glasby CJ, Timm T. Global diversity of polychaetes (Polycheata; Annelida) in freshwater. Hydrobiologia 2008; 595: 107-15.

SUBJECT INDEX

www.ingramcontent.com/pod-product-compliance
Lightning Source LLC
Chambersburg PA
CBHW041703210326
41598CB00007B/515